Signals and Communication Technology

This series is devoted to fundamentals and applications of modern methods of signal processing and cutting-edge communication technologies. The main topics are information and signal theory, acoustical signal processing, image processing and multimedia systems, mobile and wireless communications, and computer and communication networks. Volumes in the series address researchers in academia and industrial R&D departments. The series is application-oriented. The level of presentation of each individual volume, however, depends on the subject and can range from practical to scientific.

Indexing: All books in "Signals and Communication Technology" are indexed by Scopus and zbMATH

For general information about this book series, comments or suggestions, please contact Mary James at mary.james@springer.com or Ramesh Nath Premnath at ramesh.premnath@springer.com.

Jonathan Matusitz • Jayvyn Dacas

Communication in the Metaverse

Springer

Jonathan Matusitz
University of Central Florida
Orlando, FL, USA

Jayvyn Dacas
University of Central Florida
Orlando, FL, USA

ISSN 1860-4862 ISSN 1860-4870 (electronic)
Signals and Communication Technology
ISBN 978-3-031-63484-0 ISBN 978-3-031-63485-7 (eBook)
https://doi.org/10.1007/978-3-031-63485-7

This Springer imprint is published by the registered company Springer Nature Switzerland AG
The registered company address is: Gewerbestrasse 11, 6330 Cham, Switzerland

Preface

Divided into 10 chapters, this book provides a detailed account of the role of communication in the metaverse. Chapter 1 is an introduction that addresses the five objectives of this monograph: How examining the metaverse (1) enhances our understanding of human communication, (2) explains new ways of expressing one's identity, (3) shows that it is not a utopian space for communication, (4) highlights humans' creativity, and (5) poses multiple threats to humankind nonetheless. Then, the authors describe three other books that are competitors in the domain of the metaverse. Chapter 2 provides both historical and contemporary definitions of the metaverse, as well as descriptions of augmented reality (AR), virtual reality (VR), mixed reality (MR), and extended reality (XR). Then, the authors proceed to explain the three broad objectives for the development of the metaverse: (1) Using artificial intelligence (AI) to a new level, (2) mimicking offline everyday life experiences, and (3) enjoyment. What comes subsequently is an argument about the requirements for fully immersive VR, including its 12 essential features. Chapter 3 offers a general description of communication in the metaverse, with a particular emphasis on interpersonal communication, intercultural communication, and health communication. The second half of the chapter addresses the notion of social presence in the metaverse. Social presence refers to the subjective experience of communicating in one space or environment, notwithstanding the fact that one is physically "located" in another.

Chapter 4 addresses the processes of social construction of meaning, ideas about physical reality, the rules for daily communication, and our self-concept/identity in the metaverse. An important premise is that, even in VR, discourse is socially constructed and gives insight into the language used in different contexts. The end of the chapter addresses self-presentation strategies employed by metaverse inhabitants to construct their personas, including impression management strategies and the creation of avatars. Chapter 5 focuses on nonverbal communication in the metaverse, including body language, emotional expressions, facial expressions, and the voice. Of equal relevance are channels of feedback (on the part of receivers), avatar manipulation, and deep synthesis/deepfakes. The chapter ends with a discussion of emoticons and emojis.

Chapter 6 emphasizes interaction, collaboration, and training in the workplace. The main gist is that the metaverse is an environment for a vast array of joint human endeavors through work collaborations (e.g., interacting with co-workers, having business meetings), in addition to individual workloads. Certain sections highlight digital twins and workplace training for millennial employees. Chapter 7 is about education in the metaverse: immersive technologies increasingly play a large part in learning and teaching, especially with the exponential rise of AR, VR, MR, and XR hardware and applications over the past few years. A significant portion grapples with digital literacy. The end of the chapter applies Social Cognitive Theory (SCT) to learning and teaching in the metaverse.

Chapter 8 tackles deception in the metaverse. After a section defining fake news, misinformation, disinformation, conspiracy theory, and selective exposure theory, other sections cover the theory of weak ties and moral panic theory. The last part of the chapter is an account on social engineering, a term encompassing for a wide range of malicious activities undertaken through human interactions. Chapter 9 addresses other problems in the metaverse: privacy concerns, surveillance, and hate speech. For example, six categories of privacy concerns illustrate how dangerous the metaverse can be: (1) Context collapse, (2) privacy leakage during data transmission, (3) privacy leakage during data processing, (4) compromised or rogue end devices, (5) unlawful monetization of data, and (6) stalking. With respect to surveillance, a case study on China sheds light on how threatening the monitoring of public opinion can be. And targets of hate speech in the metaverse tend to be minority groups as targets.

Chapter 10 is a discussion and a list of detailed suggestions for future research. The main conclusion of this book is that communication in the metaverse rests on four principles. As such, the metaverse is (1) a complex interactive environment, (2) a socially shaped environment, (3) a controlled media environment, and (4) a disruptive innovation. The latter principle, disruptive innovation, can be broken down into four categories: disruptive communication, a higher capacity for deception, social engineering, and privacy concerns and surveillance. *Apropos* future directions for research, we offer solutions regarding communication in the workplace and reducing the digital divide.

Orlando, FL, USA

Jonathan Matusitz
Jayvyn Dacas

Contents

Chapter 1
Introduction

This book explores the diverse methods and facets of communication in the metaverse, from interpersonal relationships to health communication, and from collaboration in the workplace to social interaction in educational contexts. More specifically, the aim is to analyze the profound interplay between the metaverse and the larger communicative context in which it is entrenched. In this manner, this work concentrates on the interlaced factors underlying its origins, development, actualization, and popularity, as well as the fundamental discords, uncertainties, and dangers that can have serious consequences on both society and the individual. To date, no major volume on communication in the metaverse has been published. The few books on similar subjects focus on technical communications that are based on software or computing networks, and on strategic communications for military or defense agencies.

This book offers fresh insights on the social side of communication, covering frameworks such as social constructionism, social networking theory, social cognitive theory, and impression management—among a myriad other concepts such as interpersonal communication, intercultural communication, health communication, and social presence. In regards to the academic approaches used in this book, the authors applied multiple theories, conceptual analyses, case studies, examples, across the world, and statistics. The end of the book also includes detailed suggestions for future research.

The metaverse reflects the human aspirations to transcend the limits of space and time by enabling interactants to express their words, minds, and bodies through novel sensory experiences, representing a shared virtual universe of consensual illusion (Zhai et al., 2022). By and large, the term "metaverse" is often referred to as a digital transformation of most aspects of our lives; it can be achieved physically into a virtual approach. Other definitions of the metaverse have been suggested, including lifelogging, mutual space in virtually, embodied/spatial cyberspace, a mirror world, a setting of simulation, and collaboration. The immersive nature of the metaverse remains an essential matter because it purports to enhance our

J. Matusitz, J. Dacas, *Communication in the Metaverse*, Signals and Communication Technology, https://doi.org/10.1007/978-3-031-63485-7_1

communicative, social, and emotional intelligence abilities by imitating, or even improving, the three-dimensional scenario (Wang et al., 2022).

Our book is timely because, as the COVID-19 pandemic recomposed our daily interactions through widespread social distancing measures, the necessity for alternatives to physical and face-to-face communication greatly increased and opened new horizons through technological solutions. Although some of the latter, such as extended-reality gaming and real-time video meetings, had already existed for a long time, early 2020s technical progress and societal transformations engendered by COVID-19 accelerated the evolution of the metaverse, motivating big-tech owners and investors to pour tens of billions of dollars. All this also led to predictions that the metaverse would be a big internet battleground in the next years to come (Anderson & Rainie, 2022). A case in point is Microsoft's investment in the metaverse to the tune of US$ 70 billion in 2022, in addition to investments made by Google and Epic Games (and large corporations such as Gucci, Nike, and Walmart, to name a few). Nevertheless, these investments only represent the tip of the iceberg for a trend that has already gone far and that involves a great chunk of humankind (Caragnano, 2023).

The Five Objectives of This Book

As a *mélange* of science, technology, and social construction, the metaverse transforms the frontiers of communication that will have long-lasting impacts on society. In doing so, it illustrates how social reality changes with each phase of this digital medium. As a set of technologically driven representations (or replicas) of our three-dimensional world, the metaverse is progressively fashioning the socio-technical hypotheticals of data-based phenomena—the likely corollary being fundamental changes of important structures, processes, habits, and ways of life in the next few decades. At the heart of the systematic examination of the role of communication in the metaverse are the interrelations among humans, digital platforms, and the principles, standards, and ethics entrenched within such structures and systems (Bibri, 2022). To date, not a single book has adequately covered that breadth of knowledge. To provide an overview appropriate for students, scholars, and practitioners, the current work seeks to address five objectives: How examining the metaverse (1) enhances our understanding of human communication, (2) explains new ways of expressing one's identity, (3) shows that it is not a utopian space for communication, (4) highlights humans' creativity, and (5) poses multiple threats to humankind nonetheless.

Enhancing Our Understanding of Human Communication

The first objective of this book is to explore users' experiences in virtual reality (VR) and the metaverse at large, with the aim of generating more understanding and safer experiences for this public. As the supreme mode of human communication in the deep media sphere, the metaverse will inescapably reshape the fundamentals of such human social interactions. Hence, looking at approaches such as interpersonal relationships in the metaverse not only increases our awareness of how VR could replace face-to-face communication in the future (rightly or wrongly, and with or without the assistance of an intermediary), bit it can also reshape the concepts of space and time within human communication itself. Because it is often portrayed as an infrastructure-driven method for the betterment of society, the metaverse has the potential to promote communication at all levels: the interpersonal, social, collaborative, intercultural, and global levels, thereby becoming an overarching infrastructure for humans to enjoy such modes of communication (Jiang & Xu, 2022).

Human communication is based on experience, knowledge, character traits, and relationship skills with others. An individual's "adequate" level of communication depends on his or her aptness at knowing what is going on around him- or herself and make the right decision. Positive human relations are deemed the most important indicators of happiness and add to personal success in the real world (Dalton et al., 2010). Therefore, as the metaverse becomes more accessible to all (not just the tech-savvy ones), practitioners and scholars will have to reexamine their prior understandings of communication theory, strategies, and applications. The hard task is to ascertain the value or extent to which the metaverse will transform or, at least, expand our abilities in human communication.

Speaking of extending our abilities, Canadian scholar Marshall McLuhan's (1975) media extension theory rests on the premise that the advent of new types of media boosts our understanding of human communication by extending the span (or reach) and power of the human body. Whether it be oral, text, printing, or electronic communication, any form of communication media technology will externalize our five senses and the ability to process and remember information (Qingguang, 2011). Within the rapid expansion of digital media such as the internet, a wide assortment of media texts, audio, images, videos, icons, and memes have been incorporated into this growing interconnected communication system. As a result, users can produce an unlimited number of connections and diverse interactions through such digital media. This situation also illustrates the "decentralized" nature of the internet, thanks to all-channel networks that allow us to be nowhere and everywhere at the same time. This has not only disempowered mass communication, but it has also empowered interpersonal communication. In the twenty-first century, such integration and decentralization have been accelerated in the metaverse (Song, 2022).

The metaverse affords many academics the opportunity to move the discipline of communication to new directions. For example, from adding fresh insights to integrated communications scholarship with new perspectives on digital media studies (Hallahan, 2008), to examining the behavior and cognitive impacts of communication on metaverse users, avatars, and other publics (Tkalac, 2008), the new approaches promulgated throughout ten chapters in this book will inspire updated evaluations and definitions of current communication approaches. Having said that, academics should emphasize not only the medium used but also the interconnections between micro-level and macro-level facets of social life within virtual environments. The media and communication field has experienced an unprecedented turn of events through the changing nature and behavior of digital media audiences occasioned by the metaverse or other platforms. The long-established notion of "audience"—as mere listeners, viewers, or consumers of mass media content—is not viable anymore because the exponential broadening of media channels and sources has reconfigured audiences entirely. Put differently, the age-old collective term for media receivers (in the "simple sequential model" sense of the mass communication process) is hardly applicable in the metaverse because the latter does not operate on a "source-channel-message-receiver-effect-feedback" principle. The exponentiality of the metaverse has given users novel roles—many of which are yet to be discovered. In a nutshell, the communication process is going through unforeseen forms of communication (Obiaje & Adelabu, 2022).

Explaining New Ways of Expressing One's Identity

The relationship between one's self-identity and the manner in which people present themselves in the digital realm has been the focus of manifold studies. Yet, these have mostly looked at online blogs or social networking sites such as Facebook (see Michikyan et al., 2015; Paliszkiewicz & Madra-Sawicka, 2016). Notwithstanding the fact that the results from these studies can be used to demonstrate, in part, what transpires in the metaverse, there is still a need to understand why metaverse users project themselves in ways that align with their self-identity and the personal information they reveal to others. The idea of altering an existing identity or fashioning a new identity in VR is fairly new. This is why researchers should address how people present themselves through reembodiment in digital environments. The idea of utilizing the metaverse to transform or reconstruct one's identity through self-disclosure should continue to be explored.

In light of the significant shifts in behaviors, attitudes, and lifestyles of youths, it would be reasonable to make attempts at understanding the myriad possibilities that the metaverse provides for identity construction—and the dependence on recently developed digital platforms for identity presentation (i.e., impression management) (Zhou & Tian, 2023) and avatar formation (Lee et al., 2023). Given the promise to offer enhanced realism and presence, the metaverse can soon become a venue that meets most of the youths' identity needs, especially if they have not been met at public events or through social interactions. The metaverse can improve one's identity through virtual social support; indeed, even digital social interactions can be conducive to better psychological well-being throughout an individual's lifespan (Oh et al., 2023).

Showing It Is Not a Utopian Space for Communication

The third objective is to demonstrate that the metaverse is not a utopian space for communication. Before even delving deeply into the gist of this volume, it would be useful to ask whether or not the metaverse will become a utopian space for communication because of its gradually higher level of virtuality. The development of the metaverse is anticipated to occasion deep-rooted changes to private rooms, workplaces, and collectivities as a whole—with up-to-the-minute tools and possibilities. This will make human interactions within families, friends, and wider communities richer in social bonding. This is pivotal for safeguarding civil liberties, which have been harshly impacted by the "new normal" that was accelerated by the COVID-19 pandemic. Today, the erosion of human interactions and social bonding continues to be normalized (in spite of the alleviation of draconian measures imposed during the pandemic) and the surveillance technologies (highly present at that time) have not been removed (Bibri, 2022).

Through new ways of interacting with and reacting to each other, the metaverse will allow people to integrate practices, institutions, and policies within a social system aimed at bettering our lives and, ultimately, making us thrive. One of the concerns of a techno-utopia is that it will dampen human communication, producing a distant or divisive society as a consequence of its addiction to digital technologies and owing to the reality—if truth be told—that users have a difficult time separating their lives/identities from these technologies (even for the greater good of humanity). It is the authors' conviction that the metaverse represents a series of steps toward an arrangement of digital platforms in which interactants can effortlessly contact and interact with each other irrespective of time or location. In this process, it will reduce fragmentation between members of society, as well as among grouping of cultures and societies themselves. To this point, there is also a requirement for reexamining urban imaginaries with respect to present-day urbanism, where a great degree of daily living is digitally and virtually mediated (Bibri, 2022).

Techno-utopianism alludes to any set of beliefs and values founded on the principle that progress in science and technology will, in due course, achieve a utopian ideal or an impeccable society in the distant future. In this hypothetical situation, everyday life, in consort with processes of government, laws, and social conditions, will be exclusively made for collective benefits and the well-being of its citizenry. Detractors of techno-utopianism draw attention to its tendency to shelve the positive outcomes of laws while highlighting government interference and infringement on privacy and civil liberties. This account of illusory imaginaries of our future world as a techno-utopia summons time-honored cultural fears about the effects of socio-technological advances on society (Bibri, 2022).

Experts remain divided about the prospective success of a truly immersive metaverse. Some predict that augmented- and mixed-reality environments will turn out to be useful in our day-to-day interactions. Others express concerns that existing online problems may be exacerbated if Web 3.0 development is directed by those responsible for today's overarching web platforms (Anderson & Rainie, 2022).

However, the authors of this work maintain that the metaverse does not appear to be utopian, and thus not designed for an idealistic "universe." Rather, it is intended to be influenced by economic competition of diverse entities, which include big-tech corporations such as Apple, Meta, and Microsoft (Mac et al., 2022). Propelled by modern advances in state-of-the-art technologies such as extended reality, artificial intelligence, and blockchain, the metaverse is growing out of science fiction into an imminent reality. The degree of "communicative realism" hinges on the companies' potential to carry virtual and augmented reality tools to adequate levels of interaction or collaboration among individual users. Naturally, unbridled capitalism could be an obstacle to proper communication, but only time will tell.

Highlighting Humans' Creativity

Humans are highly complex biological, psychological, social, and technological beings. They are also highly creative beings; they have demonstrated their capacity for reaching great levels of conscious and unconscious interactions in "unchartered territories." In this manner, they have also incorporated new frameworks of meanings and rationales for their interactions, which appear more "sensical" to users and explain what drives them to engage in any type of communication in new spaces. This penchant for creativity allows us to imagine states or conditions that did not exist before. As we will see in later chapters, humans are able to find many novel ways of expressing emotions, desires, and passions—e.g., from rational to irrational, from happy to depressive, from very emotional to abstract/logical, and from intelligent to limited/stupid (Raich et al., 2021).

During the China Metaverse Summit, the year 2021 was baptized the Year of the Metaverse. Is the metaverse the hope or the hype? Is it a windfall or an illusion? There is ample talk in all sectors (Zhai et al., 2022). The metaverse is certainly not an accidental phenomenon; it is well thought-out. It is an expression of the limitless expansion of our imagination. However, counter to science fiction, social relations within the metaverse can mimic those of the real world. What transpires in the metaverse is a mirror image of the three-dimensional world, a simulation. This turns the metaverse into a real-world and perceivable space to communicate and operate (Nevelsteen, 2015).

Exposing the Threats Posed by the Metaverse

A procedural problem persists when a technology is not properly regulated, leading us to risk being overly naïve in predicting its negative impacts. The book will explain that, because the metaverse is socially constructed, it can lead, inter alia, to social engineering. In this context, social engineering consists of resorting to deceptive means of communication to create a relation with the target user in order to exploit him or her for financial, ideological, political, or even sexual gains. The wide assortment of technologies and methodologies at the hands of "social engineers" in the metaverse is colossal. Additional dangers pertain to privacy concerns, surveillance, hate speech, and bullying/mobbing.

Humans are at the forefront of a new social construct where each of them can be a target of the predicaments posed by the metaverse. Youths in particular (i.e., Generation Z and millennials) spend more time online and in VR (see Matusitz & McCormick, 2012). Millennials use it for over two-and-a-half hours a day. Generation Z spends almost three hours a day on it (*World Economic Forum*, 2019). They represent the largest volume of the population in the metaverse. Roblox, a supreme user-generated game site, is designed on the concept of "metaverse," but could undoubtedly be a bastion for ill-intentioned users. Over 80% of Roblox users are under 25 years old (*Statista*, 2021). *Fortnite*, the most popular online game platform in the world, is immensely popular among users between 14 and 24 (Wilson, 2021).

Competition with Other Works

In their book titled *From Microverse to Metaverse: Modelling the Future through Today's Virtual Worlds*, Leighton Evans et al. (2022) provide various descriptions of the metaverse, including private companies' endeavors at building all-inclusive VR empires that can metricize, commercialize, and manipulate users' actions and interactions. A great deal of the book is also devoted to traditional concepts of the digital, such as social worlds, gaming worlds, user-generated worlds, and commerce/entertainment. Of particular interest in their work is the notion of microverse. Not to be confused with the Microverse® (the online school for remote software developers), microverses are the multiple social networks and spaces in the metaverse that have substituted our numerous day-to-day social functions and activities for interactional purposes. Although *From Microverse to Metaverse* was well written and adequately describes the metaverse as an experiential and virtual world that can be embraced by the vast majority of users, it contains two major flaws: (1) aside from a few chapters (e.g., *Gaming Worlds*), a significant number of sources are outdated and (2) the emphasis of the book is too broad, ranging from fitness to desire, and from social to commerce. In other words, there needs to be a stronger common denominator across all chapters.

In *Unconscious Networks: Philosophy, Psychoanalysis, and Artificial Intelligence*, Possati (2023) explains the role of communication in artificial intelligence. At first, we were pleased with the book because it deals with social network theory, human–machine interaction (i.e., what Possati calls "humans and nonhuman actants"), and the ways in which human communication unconsciously shapes technology (and vice versa). However, we quickly grew disappointed because the work turned out to be a volume written for academics and experts in the disciplines of technology studies, artificial intelligence, psychoanalysis, and philosophy.

Unconscious Networks did not focus on an important aspect of unconscious communication: nonverbal communication (e.g., facial expressions, microexpressions, and so forth). In line with these contentions, very little space was devoted to actual interpersonal relationships in the metaverse.

In the longest and most comprehensive of all three books, *Making a Metaverse That Matters: From Snow Crash & Second Life to A Virtual World Worth Fighting For*, Wagner James Au (2023) delves into both communicative and noncommunicative areas. The communicative areas emphasize community building, connections across generations, avatar formation, and collaboration in gaming. We also appreciate Au's attention to postmodernism and the dangers of an unfettered metaverse. However, to our dismay, essential communication concepts were missing. Examples include identity creation (which would be highly important when discussing community building and avatars) and nonverbal communication. Of equal disappointment is the fact that a great deal of content is dedicated to addressing the myths about the metaverse. To this point, then, the title of Au's book is a misnomer. It could be rephrased as *Truths and Myths about the Metaverse*.

References

Anderson, J., & Rainie, L. (2022). *The Metaverse in 2040*. Pew Research Center.

Au, W. J. (2023). *Making a metaverse that matters: From snow crash & second life to a virtual world worth fighting for*. Wiley.

Bibri, S. E. (2022). The social shaping of the metaverse as an alternative to the imaginaries of data-driven smart cities: A study in science, technology, and society. *Smart Cities, 5*(3), 832–874. https://doi.org/10.3390/smartcities5030043

Caragnano, R. (2023). Labour law: New workplaces in the metaverse and opportunities for cultural and heritage professions. *Athens Journal of Law, 9*(2), 211–228.

Dalton, M., Hoyle, D. G., & Watts, M. W. (2010). *Human relations*. Cengage Learning.

Evans, L., Frith, J., & Saker, M. (2022). *From microverse to metaverse: Modelling the future through today's virtual worlds*. Emerald Publishing.

Hallahan, K. (2008). Integrated communication: Implications for public relations beyond excellence. In E. L. Toth (Ed.), *The future of excellence in public relations and communication management* (pp. 299–336). Routledge.

Jiang, C., & Xu, J. (2022). Exploration of short video media communication based in the metaverse. In L.-J. Zhang (Ed.), *Metaverse—METAVERSE 2022* (pp. 18–28). Springer.

Lee, H.-W., Chang, K., Uhm, J.-P., & Owiro, E. (2023). How avatar identification affects enjoyment in the metaverse: The roles of avatar customization and social engagement. *Cyberpsychology, Behavior, and Social Networking, 26*(4), 255–262. https://doi.org/10.1089/cyber.2022.0257

Mac, R., Frenkel, S., & Roose, K. (2022, October 9). Skepticism, confusion, frustration: Inside Mark Zuckerberg's metaverse struggles. *The New York Times*, A1. Retrieved on May 19, 2023 from https://www.nytimes.com/2022/10/09/technology/meta-zuckerberg-metaverse.html

Matusitz, J., & McCormick, J. (2012). Sedentarism: The effects of internet use on human obesity in the United States. *Social Work in Public Health, 27*(3), 250–269. https://doi.org/10.1080/19371918.2011.542998

McLuhan, M. (1975). *Understanding media: The extension of man*. Routledge.

Michikyan, M., Dennis, J., & Subrahmanyam, K. (2015). Can you guess who I am? Real, ideal, and false self-presentation on Facebook among emerging adults. *Emerging Adulthood, 3*(1), 55–64. https://doi.org/10.1177/2167696814532442

Nevelsteen, K. J. L. (2015). Virtual world, defined from a technological perspective and applied to video games, mixed reality, and the metaverse. *Computer Animation and Virtual Worlds, 29*(1), 1–36. https://doi.org/10.48550/arXiv.1511.08464

Obiaje, K. M., & Adelabu, O. T. (2022). Media audience emergence as forces to be reckoned with in the internet age. *Sapientia Global Journal of Arts, Humanities and Development Studies, 5*(1), 101–111.

Oh, H. J., Kim, J., Chang, J., Park, N., & Lee, S. (2023). Social benefits of living in the metaverse: The relationships among social presence, supportive interaction, social self-efficacy, and feelings of loneliness. *Computers in Human Behavior, 139*(6). https://doi.org/10.1016/j.chb.2022.107498

Paliszkiewicz, J., & Madra-Sawicka, M. (2016). Impression management in social media: The example of LinkedIn. *Management, 11*(3), 203–212.

Possati, L. M. (2023). *Unconscious Networks: Philosophy, psychoanalysis, and artificial intelligence*. Routledge.

Qingguang, G. (2011). *A course for communication studies*. China Renmin University Press.

Raich, M., Dolan, S. L., Ulrich, D., & Cisullo, C. (2021). Human uniqueness at the Dawn of intelligent machines. *The European Business Review, 3*, 39–59.

Song, Z. (2022). Interpersonal communication research in metaverse—Taking Si-Fi films as example. *Advances in Social Science, Education and Humanities Research, 664*, 2377–2382.

Statista. (2021). Roblox Gamers Users Distribution Worldwide September 2020, by Age. Retrieved on October 4, 2022 from https://www.statista.com/statistics/1190869/roblox-games-users-global-distribution-age/

Tkalac, A. (2008). The application of situational theory in Croatia. In E. L. Toth (Ed.), *The future of excellence in public relations and communication management* (pp. 527–543). Routledge.

Wang, S.-M., Yaqin, M. A., & Hsu, F.-H. (2022). Improving emotional intelligence in the new normal using metaverse applications for digital native. In M. Kurosu (Ed.), *Human-computer interaction theoretical approaches and design methods* (pp. 527–541). Springer.

Wilson, S. (2021). Where Brands Are Reaching Gen Z. *Harvard Business Review.* Retrieved on October 4, 2022 from https://hbr.org/2021/03/where-brands-are-reaching-gen-z

World Economic Forum. (2019). This graph tells us who's using social media the most. Retrieved on October 4, 2022 from https://www.weforum.org/agenda/2019/10/social-media-use-by-generation/

Zhai, X., Chu, X., Wang, M., Zhang, Z., & Dong, Y. (2022). Education metaverse: Innovations and challenges of the new generation of internet education formats. *Meta, 3*(1), 1–13. https://doi.org/10.54517/met.v3i1.1804

Zhou, B., & Tian, Y. (2023). Network matters: An examination of the direct and mediated influences of network size and heterogeneity on WeChat fatigue. *Computers in Human Behavior, 139*, 107489. https://doi.org/10.1016/j.chb.2022.107489

Chapter 2
What Is the Metaverse?

The word "metaverse" was seen for the first time in *Snow Crash*, a science-fiction novel written by Neal Stephenson (1992). Stephenson pictured the human race as one escaping the callous reality of a collapsed economy that had besieged the world. So, the human species moved into a virtual environment. This mirrors other authors' appropriation of futuristic scenarios mingled with contemporary histories. Examples include works by Isaac Asimov (1920–92), Arthur C. Clarke (1917–2008), Ray Bradbury (1920–2012), H. G. Wells (1866–1946), Jules Verne (1828–1905), and even the traditional science fiction of the Hanna-Barbera animation studio with its *Jetsons* (1962–63), particularly its extravagant robotic contrivances, aliens, holograms, and quirky inventions. Stephenson's novel—and his few others too—includes concepts and ideas relevant to this topic, such as "headsets" and "goggles" that allowed protagonists to engross themselves into a fictional virtual world before goggles and technologies even existed—inspiring the global market for these products to make profitable outcomes. These concepts relate to the notion of extended reality (XR) proposed by Paul Milgram—a comprehensive term for immersive technologies such as augmented reality (AR), mixed reality (MR), virtual reality (VR), and 360-degree interactive reality (Bibri et al., 2022). The next section provides contemporary definitions of the metaverse.

Definitions of the Metaverse

The metaverse denotes a concept of an enduring digital world that merges technologies (such as the aforementioned AR, VR, and XR) with physical reality and supplants the present-day internet (Sykownik et al., 2022). It is a new computer-mediated environment that comprises virtual "worlds" in which people can interact with each other directly or through avatars. An avatar is an icon or figure representing a particular person in an online space (Miao et al., 2022). From this vantage point, the

J. Matusitz, J. Dacas, *Communication in the Metaverse*, Signals and Communication Technology, https://doi.org/10.1007/978-3-031-63485-7_2

metaverse heralds the current apex of globalization—an interconnection between people, ideas, and technologies (Matusitz, 2021). The metaverse includes a definite set of hardware (i.e., head-mounted virtual-reality devices, or headsets) that serves as its core access and use, together with unique operating systems (e.g., Meta/Oculus) and apps that provide those virtual "worlds" (e.g., Altspace by Microsoft, Horizon by Meta). The metaverse is an intrinsically social place that offers room for a wide continuum of collective human activities, from entertainment (e.g., gaming, watching movies, and even bowling) to work collaborations (e.g., interacting with colleagues, holding business meetings), besides personal needs (Hennig-Thurau et al., 2022).

No single definition of the metaverse is available out there, as the concept constantly needs streamlining and development, thereby leading major players such as Meta (i.e., Facebook) to seize the opportunity to provide their own definitions and descriptions of what it could be (*BBC News*, 2021). From an abstract standpoint, the metaverse is a virtual space that can be accessed through devices such as head-mounted displays in which users interact with others as if they were physically present, rendering the interaction as natural as possible (including body language as well). The main idea is to devise a fully immersive environment where people use avatars to communicate and, by the same token, to imitate human interaction in the real world (Alvim, 2022). Until recent years, VR technology was predominantly associated with the gaming industry where players cooked up their own virtual worlds such as *Fortnite*. *Fortnite* recognizes the promise of the metaverse and has changed its own basic conception of it—from the state of being a game to being a social place for metaverse users. *Fortnite* also incorporates additional features in which users can spend money to improve their avatars, such as embodying aesthetic characteristics like dressing up their characters (Herrman & Browning, 2021).

Indeed, the technologies that corporations allude to when describing "the metaverse" tend to include VR—typified by unshakable virtual worlds that are continue to function when one is asleep—and AR (that mingles features of the digital and physical worlds). However, this does not mean that the metaverse is only accessible through VR or AR. A significant number of virtual environments—such as parts of *Fortnite* that can be logged on to through PCs, game consoles, and smartphones—plainly refer to themselves as "the metaverse" (Ravenscraft, 2022). Attention to the metaverse skyrocketed in 2021–22, mostly driven by Facebook's decision to rebrand itself as "Meta." At the present time, the metaverse is commonly composed of immersive extended reality (XR) spaces in which communication takes place among people and/or automated entities. Whereas some are routine interactions with AR apps on computers and phones, others are interactions in the more immersive milieus of gaming and fantasy worlds. And additional types of interactions take place in "mirror worlds" that replicate real-life environments (Anderson & Rainie, 2022).

Etymology

By and large, "metaverse" is a term that combines *meta*, meaning virtual, and *universe*, meaning worldview. It is a technology that has grown exponentially as a result of the continued emphasis on online activity (Mystakidis, 2022). Because of its *universe* aspect, the metaverse can be said to be a universal immersive world that links (1) hardware technologies that make up the virtual world, (2) the interactants using these technologies, and (3) the associated three-dimensional (3D) world. This is how the metaverse connects interactants in the real and virtual worlds. Put another way, the metaverse is a set of technologies that allow users to have an interactive encounter in a specific virtual reality (Hwang et al., 2022).

Let us look at the concept of "universe" even further. From the Latin *universus*, it translates as "all together," "all in one," "whole," or "entire." *Universum* represents "all things," "everyone," or "the whole world." In modern English, it signifies "the sum of everything that exists in the cosmos, including space and time," that is, physical reality as a whole. It may refer to "all conceivable instances of a thing" (e.g., a universe of possibilities) (Dolata & Schwabe, 2023). The prefix *meta* ("among," "with," or "after") is an Ancient Greek term. It has been used primarily in the word metaphysics (*meta ta physika*) or "after the things of nature." Metaphysics is the discipline of what comes "after," "beyond," or "above the physical world." The metaverse, then, goes beyond or after the current version of the internet. Gradually, the focus of metaphysics has evolved into exploring ontology (the study of being/existence), including the links between mind and matter, substance and attribute, or potentiality and actuality. Taken as a whole, *meta* can help explain new meanings such as "transcending" and "a level above." It can also include the definition of "distinct from but about other things of the same type" or "self-referential" (Dolata & Schwabe, 2023, pp. 239–66). Within the context of this work, this is exactly what the metaverse exemplifies.

AR, VR, MR, and XR

Augmented reality (AR) is a comprehensive term for technologies that augment or modify the current reality (Riar et al., 2021), whereas in virtual reality (VR) the sensory information and stimulus of the real world tends to be blocked and subdued (Manis & Choi, 2019). Hence, while VR is deemed a digital technology of choice for replacing perceived reality (Xi & Hamari, 2021), AR and VR give distinct forms of experiences to users (Fromm et al., 2023). With respect to AR, the "augmenting" information and content can bestow users with interactivity, intensity, power, and innovation (Yim et al., 2017). On the other hand, VR is deemed to generate immersiveness (Suh & Prophet, 2018), telepresence (Lee & Chung, 2008), and a sensation of "being there" (IJsselsteijn & Riva, 2003). Mixed reality (MR) is a *mélange* of physical and digital realms, releasing natural and instinctual 3D human, computer, and environmental interactions (Marr, 2021). It is a distinct but complementary virtual experience highly correlated with the real world (Edwards et al., 2018).

Sometimes called cross reality, extended reality (XR) denotes all sorts of immersive technologies, their corresponding human–machine interactions, and environments such as the metaverse that depend on these and the utilization of spatial computing for optimal actions in these systems. XR technology combines the physical and digital worlds and mixes the aforementioned concepts of AR, VR, and MR as well (Chuah, 2019). The evolution of immersive technologies began in the 1960s when the first headset was patented by a researcher named Ivan Sutherland. Today, the sensitivity of data obtained by XR applications includes users' physical appearance and biometric data (e.g., iris identification, pupil dilation, and gaze tracking). The acquired data can give out specifics of users such as their social life, lifestyle, hobbies, financial resources, home interior, and profession (Qamar et al., 2023).

As early as 1975, *Myron Krueger* put a "videoplace" into practice, which consisted of integrating video cameras and projectors to construct a VR world. This experiment led to rapid development and various other XR applications (Qamar et al., 2023). Many sensors and displays were soon installed on XR devices to examine sensors' inputs and produce congruous reactions to the display as output for an immersive experience. XR devices are built to create emotions and feelings so that they can be seized through sensory systems attached to the human body—giving the actualization that the user is *there* in an immersive world. Vibro-haptic and thermo-haptic sensors inherent in these devices function by producing vibrational and thermal sensations in the virtual world. One of the results, for example, is that the user's skin can now sense physical touch (Ko & Rogers, 2021). As one can see, in the metaverse, consumers' encounters with digital entities work through XR's increasing sophisticated computer graphics and expanding bandwidth that enable the realization of world-like objects and situations. XR can clarify and improve such usability, performance, efficacity, and productivity of interactions in the metaverse at a fast pace (Muñoz-Saavedra et al., 2020).

Objectives of the Metaverse

The metaverse is made up of infrastructures, arrangements, and standards that are linked to the physical world through intermediaries (without being separated from it). Because of the large volume of content generated in this digital realm, one key objective of the metaverse is to deliver massive-scale and perpetual content production for both the physical and virtual environments. As a result of a multitude of digital technologies for speech recognition, machine vision, and human language processing, the metaverse grows into a simpler-to-use and more practical platform. Digital avatars provide visual dialogue situations using artificial intelligence (AI) and virtual reality. Digital twins—virtual models conceived to accurately mirror physical objects—allow users to gather and analyze a great amount of data, affording a 360-degree view of the devices. By extension, the metaverse proves to be useful in the domains of education and training (Pooyandeh et al., 2022). Three broad objectives for the development of the metaverse have been identified: (1) using artificial intelligence (AI) to a new level, (2) mimicking offline everyday life experiences, and (3) enjoyment.

Using Artificial Intelligence (AI) to a New Level

This first objective is the broadest and can be turned into an entire book or volume in and of itself. Artificial intelligence (AI) is generally described as the framework and development of computer systems to complete tasks ordinarily done through human intelligence; examples include visual and speech recognition, decision-making, and language translation. AI is also used to examine user behaviors in much more detail and scrutinize data sprung out of user interactions and digital twins to guarantee the continuation of the virtual world. Many institutions and organizations have established AI guidelines for frameworks, bulk data, and use cases (Kang et al., 2023).

By merging the physical and virtual worlds, the metaverse allows AI to take part in a much broader range of undertakings in connection with creation, display, leisure, social networking, and marketing—among many other activities. AI is increasingly being used to validate, populate, and preserve the metaverse. In addition to using AR, VR, XR, the metaverse is built on blockchains (shared databases distinct from regular databases in their storage of information) and other types of digital networks. In this ever-expanding virtual environment, AI will arguably manufacture a digital world in which any user can—in theory—guardedly and freely participate in social and economic endeavors that transcend the physical realm (Sivasankar, 2022). As a result of AI, people can benefit from new forms of learning, be assisted by virtual teaching aides, and enroll in educational courses with the same quality as those in brick-and-mortar locations. The future shows signs of bountiful opportunities for partnerships and alliances on the AI-driven metaverse (Pooyandeh et al., 2022).

A specific illustration of this can be seen in data learning. It is conventional wisdom that learning data is a pivotal component of AI and machine learning (ML). ML is a practical application to forecast bandwidth and views, as well as boost video streaming bitrate to enhance the quality of experience (QoE). When a model is fed archival data, it acquires and memorizes the prior model's outputs and, consequently, proposes new ones based on them (improving QoE in the process). The higher the volume of data and human feedback fed into the model, the higher the number of quality outputs during each operation/application. This increases the likelihood that AI will perchance be able to fulfill tasks and produce accurate results better than what humans do. In this fashion, there will be lower incidences of human interference, and the scalability of the metaverse will grow over time (Ruan & Xie, 2021).

Another example of the power of AI in the metaverse is instinctive interface. In this case, AI can assist with human–computer communication. When putting on an AI-enabled VR headset, the user can detect and predict his or her electrical and muscle patterns (through the headset sensors) to establish how he or she wants to move in the metaverse. Thus, AI can easily recreate a detailed and minute feeling of touch. It is also beneficial for voice-enabled navigation, allowing users to interface with virtual items without hand controls. This phenomenon can be chalked up to the term "user actions." User actions are behaviors performed by users via various sensory channels, such as gaze and gestures through body language channels—which are the input of an interactive function (Zhao et al., 2022).

Lastly, enter a recent pinnacle of AI today: digital wallets. Digital wallets are software applications that store and deal with cryptocurrencies or assets. Cryptocurrency is a popular digital asset of which the development is commensurate with advances in financial technology, and it offers significant research opportunities. Each wallet can operate through a different software application to gain access to blockchain data and run on them. Any wallet can be used for one or more cryptocurrencies and applications. Public addresses are essentially cryptocurrency final account numbers; the purpose is to acquire a particular form of cryptocurrency and it can be traded publicly. The digital wallet helps retrieve any transaction linked to this address in the corresponding blockchain. In addition to viewing and accessing the balances associated with an address, the wallet can move money on the blockchain until the user owns the address. In the absence of a digital wallet, blockchain transactions cannot be effectuated. In regards to security, this mostly depends on the type of device/mode used, such as desktop, smartphone, internet, paper, and hardware. Digital wallets are also more susceptible to security threats than offline wallets (Ersoy & Gürfidan, 2022).

It goes without saying that digital wallets must run on a direct internet connection. This makes them very facile to utilize for moving money around. Unfortunately, cyber-attackers can take advantage of software vulnerabilities, through malware or phishing. It is not unusual for digital wallets to remain disconnected from the network, in which case they would be used through a specific hardware device or a paper wallet (for which the private key is printed on paper and "hidden" in a safe location). This is termed a "cold wallet," which is never to be found online unless the user wants to turn it into digital assets (Das et al., 2019).

Mimicking Offline Everyday Life Experiences

The metaverse is growing in popularity on account of virtual environments in which users can communicate and experience deep interactivity, mimicking offline daily life experiences (Hackl et al., 2022). This goes beyond the simple convergence of the physical and virtual worlds, as described in the previous section. In fact, it is essential to be cognizant of the perceived absence of distinction between the offline world and real world, where relationships and behaviors are experienced as if they were real. The genuineness of these experiences has been ascertained in past research and will be increasingly been made more powerful through future VR technologies (Bailenson et al., 2004; Bailenson & Yee, 2005). This mimicking of the offline, everyday life experiences also demonstrates that Web 2.0 is gradually evolving into its new form, Web 3.0, except that its newest version is having a decentralized layer sitting atop the existing online infrastructure. Explained differently, through this new iteration of the web, a human-to-human network will facilitate new forms of digital wallets, trade currencies, and stored digital data without fees or terms and conditions set by institutions (Boder, 2021).

Early forms of the metaverse were available through PC access and had less consistency because of time and space constraints. Now that we are moving into Web 3.0, users can easily access the metaverse at any time and location because mobile devices can connect Wi-Fi. Roblox, the supreme user-generated game site, is often played on mobile devices and contains 50 million games—one of the largest in the world (*Statista*, 2021)—and the total monthly usage time exceeds three billion hours. Individuals spend more time on the metaverse than on simple social networking sites (e.g., TikTok, YouTube). The metaverse has a virtuous cycle ecosystem where the inflow and income of producers climb in proportion to the percentage of users and usage time—at the same time, it offers other various contents. Therefore, sales of digital advertisements climb as well. Similarly, the current metaverse diverges from the previous one in that program coding can be directly achieved in the metaverse itself. Fast and direct program coding allows the metaverse to become larger with abundant social meanings (e.g., fashion, entertainment, education, and work) based on immersive interactions (Park & Kim, 2022).

Under these circumstances, Web 3.0 operates on the principle that online and offline selves do not differ. Through the technological advancement of deep learning-based high-precision recognition models (and natural generation ones), the metaverse can be bolstered by a multiplicity of factors, from easy smart phone-based access to connectivity with reality (using digital currency). This consolidation of enhanced social activities and neural-net methods implies that the metaverse is suitable for the present, unlike the previous version of it (Park & Kim, 2022).

Enjoyment

Enjoyment in a particular environment is a use motive (Kaur et al., 2020). Enjoyment is a user's intrinsic behavioral motive for relying on novel technologies such as interactive game media, facilitating the determination of intention (Lee & Cho, 2020). In a virtual realm, enjoyment is primordial to the determination of users' virtual reality experience behavior, and users should secure enjoyment through the entertainment factors and innovative technologies inherent to the virtual world. Users have expressed contentment with their metaverse experiences not only through interactive games but also through 3D content in general. The latter, in and of itself, is a massive motivator for the use of the metaverse (Kim & Lee, 2022). And this can be achieved in numerous immersive ways (which will be discussed throughout this volume). Most users in this category are youths between 13 and 18 years of age. In their eyes, the metaverse is a substitute for communicative engagement, education, work, pleasure, and immersive play in the three-dimensional world (Maloney, 2021).

All things considered, it would be difficult to discuss enjoyment without discussing avatar creation. Through avatar creation, users find themselves in the centerpiece of the metaverse because the more "accurate" one's avatar, the better the quality of experience for the user. A simple AI-based avatar system may include 2D-user photos or 3D-scans to produce a highly realistic virtual reproduction. It then plugs in a wide spectrum of facial expressions, emotional states, hairstyles, and age-appropriate traits to render the avatar more dynamic. An illustration of this scenario was witnessed in "Ready Player Me" when the world's leading avatar platform helped users create their own avatars in the metaverse. Meta is constantly working on its own version of it (Sivasankar, 2022).

In futuristic scenarios, enjoyment in the metaverse is portrayed as a hypothetical version of the internet that looks like a single, universal, and immersive environment facilitated by the simple use of VR and AR headsets (O'Brien & Chan, 2021). In present-day situations, however, the metaverse can be just that or it can be a much more complex network of 3D-virtual worlds based on social connections (Clark, 2021). These offer a broad range of capabilities to allow communication and information sharing in more interactive ways than those in two-dimensional situations. At best, advanced VR systems can completely immerse people in a virtual environment, where they interface with both virtual objects and individuals as if they were in real time. This phenomenon lends itself to hands-on experience that either mimics real-world experiences or affords the storing and exchanging of complex information in unprecedented ways (Dick, 2021).

Requirements for Fully Immersive VR

A central feature of the metaverse is that it encompasses open and shared VR spaces (which we like to call "worlds") that are computer-based, interconnected, immersive, and participatory (Uspenski & Guga, 2022). It has been established that VR is a digital, immersive three-dimensional visualization world in which users can interact (Pesce, 1995). It is composed of a quasi-infinite number of interconnected data structures linked through electronic codes. The codes are "laid" or "kept" on interconnected servers that stockpile both short- and long-term memory. The connective capabilities of the metaverse will fulfill the needs of many users when they reach the level of noospheric intelligence (i.e., which has to do with superhuman consciousness and mental activity). Nevertheless, at the time of this writing, its feasibility can still provide a wide array of independent intelligences and a "wisdom of crowds," as its plasticity is in constant progress (Vita-More, 2010).

VR empowers users to interact play, learn, work, and collaborate in such 3D virtual environment with an immersive participation, by creating an imaginary environment analogous to the real world, imbued with "authentic" sounds, images, and other sensations. The metaverse can mature even further by reaching higher levels of fully immersive experiences, large numbers of concurrent users, and seamless connectivity. This, of course, presents extraordinary challenges to the sixth generation (6G) wireless system, including omnipresent connectivity, ultra-low latency, supremely high capacity and reliability, and tougher security. To the point made about large numbers of concurrent users, to meet requirements for immersive and trouble-free experiences of mass users, full coverage sensing, smooth and coherent computation, reliable caching, and strong security should be carefully considered to move into future 6G systems (Tang et al., 2022).

VR operates on 12 essential features. The list below is based, in part, on Baym's (2010) work:

1. *Interactivity*: It is a mode of communication that can be accessed remotely (i.e., with an internet connection) and concurrently by a large number of people, with the command inputs of one or few individuals affecting the command results of other users. Interactivity denotes social interactivity, which is "the ability of a medium to enable social interaction between groups or individuals" (Fornäs et al., 2002, p. 23). Liu and Shrum (2002) also define it as "the degree to which two or more communication parties can act on each other, on the communication medium, and on the messages and the degree to which such influences are synchronize" (p. 54). Social media platforms are designed particularly to promote interactivity by creating easy and effective connections.
2. *Physicality*: Users access VR through an interface that mimics a first-person physical environment on a computer screen; the environment is usually regulated by the natural laws of Earth and is typified by an abundance of resources (Real de Asúa et al., 2022).
3. *Replicability*: The long-lasting feature of our online sociality allows interactions to be captured and conveyed to intended (or unintended) recipients as a central differentiating feature of digital communications (Donath & Boyd, 2004; Papacharissi, 2010).
4. *Persistence*: A VR platform constantly runs, regardless of whether or not users are on it. It remembers the location of users or objects, as well as the ownership of the latter (Real de Asúa et al., 2022).
5. *Temporal structure*: This feature denotes the degree to which communications are synchronous or asynchronous. The strengths of synchronous communication are its power to significantly reduce the spatiotemporal distance and to generate quasi-instantaneous responses. On the other hand, asynchronous communication trades immediacy for the power to enable large groups of users to interact efficiently; for instance, based on their schedules or time zones. Unlike face-to-face communication, asynchronicity also affords users the opportunity to contemplate their contributions before sending them, as well as to handle their self-image more cleverly (Baym, 2010).

6. *Social cues*: Face-to-face or, as Fortunati (2005) contends, body-to-body, communications is steep in visual, audio, and contextual signals. In VR environments, these cues tend to decrease. With fewer verbal and nonverbal cues available in social encounters, users' digital social interactivity does not measure up because they run the risk of being understood (Walther et al., 2005).
7. *Storability*: The storability of data and activities stands in sharp contrast to the ephemeral character of face-to-face interactions and provides a new outlook on how people perceive public and private spheres (Beer, 2008; Ellison et al., 2012).
8. *Reachability*: The size of the public that can be reached on VR platforms can be astronomical. As opposed to face-to-face communication, online interactions are not confined to the volume of people who can be physically present in the space. This is because digital discourse not only travels quickly but it also travels widely as well (Gurak, 2001).
9. *Mobility*: The degree to which the VR platform is portable is quasi-unlimited. This feature is increasingly important because, as such technologies improve, users can remain socially connected. The "connected anywhere" quality of VR enhances users' autonomy and accountability—e.g., through location tagging. This, in turn, affects social interactions (Schegloff, 2002). As an alternative to geographically limited face-to-face communications, mobile internet-connected devices allow users' VR platforms to never use more than a few keystrokes during each connection. According to Facebook statistics, about one billion users access Facebook through mobile service operators (Facebook) (Baym, 2010).
10. *Social surveillance*: This feature consists of tracking the actions, passions, and beliefs of VR platform members (Duffy & Chan, 2019).
11. *Social searching*: This feature consists of checking the status of existing contacts. O'Riordan et al. (2016) explain that VR users such as profile building, social connectivity and interactivity, content discovery, information sharing, and content aggregation.
12. *Social browsing*: This feature consists of establishing new connections (Lampe et al., 2006).

The digital and the physical worlds are becoming increasingly identical to each other. Nevertheless, the 12 essential features described above reflect the outlook of the metaverse today and, as time passes, would constantly need to be enhanced or supplemented by other essential features. To date, the two worlds remain distinguishable from each other because they are ontologically, symbolically, conceptually, and semiotically distinct. In the event that they become identical, the difference between "cyber" space and "physical" space will no longer be relevant. Bown (2017), like Georges (2009), uses "real" when we use "physical," and adds this comment: "The focus here is not on how assumptions about objectivity dictate our relations to the electronic object but on how our relationship with electronic objects dictate our consumption of 'real' ones" (Bown, 2017, p. 23). In this context, Bown proposes that the digital and physical are distinguishable, notwithstanding with commonalties and transmediality of meaning and message (Davidson, 2022).

Nevertheless, although the digital and the physical worlds remain different from each other, they should not be looked at as two distinct variables of the creation of space. They are highly correlated, and none is dominating the other. Rather, they are complementing each other. With the metaverse, such discussions expand into covering socio-spatiality as a method of intertwining virtual and physical social connections and interactions—or phygital (i.e., physical–digital) experiences—applied to the foggy boundaries and spatialities of three-dimensional space and virtual space (or phygital space). And all this occurs under the dominion of socio-technological factors (Hamurcu, 2022; Hamurcu & Terzi, 2022).

References

Alvim, L. (2022, January 28). How the metaverse could impact the world and the future of technology. *ABC News*. Retrieved on September 20, 2022 from https://abcnews.go.com/Technology/metaverse-impact-world-future-technology/story?id=82519587

Anderson, J., & Rainie, L. (2022). *The metaverse in 2040*. Pew Research Center.

Bailenson, J. N., & Yee, N. (2005). Digital chameleons: Automatic assimilation of nonverbal gestures in immersive virtual environments. *Psychological Science, 16*(10), 814–819. https://doi.org/10.1111/j.1467-9280.2005.01619.x

Bailenson, J. N., Beall, A. C., Loomis, J., Blascovich, J., & Turk, M. (2004). Transformed social interaction: Decoupling representation from behavior and form in collaborative virtual environments. *Presence: Teleoperators & Virtual Environments, 13*(4), 428–441. https://doi.org/10.1162/1054746041944803

Baym, N. (2010). *Personal connections in the digital age*. Polity.

BBC News. (2021, October 18) Apparently, It's the next big thing. What is the metaverse? *BBC News*. Retrieved on September 20, 2022 from https://www.bbc.com/news/technology-58749529

Beer, D. (2008). Social network(ing) sites... Revisiting the story so far: A response to Danah Boyd and Nicole Ellison. *Journal of Computer Mediated Communication, 13*(2), 516–529. https://doi.org/10.1111/j.1083-6101.2008.00408.x

Bibri, S. E., Allam, Z., & Krogstie, J. (2022). The Metaverse as a virtual form of data-driven smart urbanism: On post-pandemic governance through the prism of the logic of surveillance capitalism. *Computational Urban Science, 2*, 24. https://doi.org/10.1007/s43762-022-00051-0

Boder, R. (2021, November 17). Entrepreneurs should embrace Web 3.0. *Entrepreneur*. Retrieved on February 23, 2023 from https://www.entrepreneur.com/science-technology/entrepreneurs-should-embrace-web-30/394083

Bown, A. (2017). *The playstation dreamworld*. Polity Press.

Chuah, S. H.-W. (2019). Wearable XR-technology: Literature review, conceptual framework and future research directions. *International Journal of Technology Marketing, 13*(3), 205–259. https://doi.org/10.1504/IJTMKT.2019.104586

Clark, P. A. (2021, November 15). The metaverse has already arrived. Here's what that actually means. *Time*. Retrieved on August 24, 2022 from https://time.com/6116826/what-is-the-metaverse/

Das, P., Faust, S., & Loss, J. (2019). A formal treatment of deterministic wallets. *Proceedings of the 2019 ACM SIGSAC Conference on Computer and Communications Security, 8*, 651–668.

Davidson, K. J. (2022). *Semiotic modelling of identity and communication in virtual reality, augmented reality, and mixed reality*. University of Tartu.

Dick, E. (2021). *The promise of immersive learning: Augmented and virtual reality's potential in education*. Information Technology & Innovation Foundation.

Dolata, M., & Schwabe, G. (2023). What is the metaverse and who seeks to define it? Mapping the site of social construction. *Journal of Information Technology, 38*(3), 239–266. https://doi.org/10.1177/02683962231159927
Donath, J., & Boyd, D. (2004). Public displays of connection. *BT Technology Journal, 22*(4), 71–82. https://doi.org/10.1023/B:BTTJ.0000047585.06264.CC
Duffy, B. E., & Chan, N. K. (2019). 'You never really know who's looking': Imagined surveillance across social media platforms. *New Media & Society, 21*(1), 119–138. https://doi.org/10.1177/1461444818791318
Edwards, A., Edwards, C., Wahl, S. T., & Myers, S. A. (2018). *The communication age: Connecting and engaging* (3rd ed.). Sage.
Ellison, N. B., Hancock, J. T., & Toma, C. L. (2012). Profile as promise: A framework for conceptualizing veracity in online dating self-presentations. *New Media & Society, 14*(1), 45–62. https://doi.org/10.1177/1461444811410395
Ersoy, M., & Gürfidan, R. (2022). Blockchain-based asset storage and service mechanism to metaverse universe: Metarepo. *Transactions on Emerging Telecommunications, 34*(3). https://doi.org/10.1002/ett.4658
Fornäs, J., Klein, K., Ladendorf, M., Sundén, J., Sveningsson, M., & Borderlands, D. (2002). *Cultural studies of identity and interactivity on the internet*. Peter Lang.
Fortunati, L. (2005). Is body-to-body communication still the protype? *The Information Society, 21*(1), 53–61. https://doi.org/10.1080/01972240590895919
Fromm, J., Eyilmez, K., Baßfeld, M., Majchrzak, T. A., & Stieglitz, S. (2023). Social media data in an augmented reality system for situation awareness support in emergency control rooms. *Information Systems Frontiers, 25*(3), 303–326. https://doi.org/10.1007/s10796-020-10101-9
Georges, F. (2009). Self-representation and digital identity: A semiotic and quali-quantitative approach to the cultural empowerment of the Web 2.0. *Réseaux 2, 154*(2), 165–193.
Gurak, L. J. (2001). *Cyberliteracy: Navigating the internet with awareness*. Yale University Press.
Hackl, C., Lueth, D., & Di Bartolo, T. (2022). *Navigating the metaverse: A guide to limitless possibilities in a Web 3.0 world*. Wiley.
Hamurcu, A. U. (2022). The metaverse, online communities, and (real) urban space. *Urbani izziv, 33*(2), 73–81. https://doi.org/10.5379/urbani-izziv-en-2022-33-02-01
Hamurcu, A. U., & Terzi, F. (2022). Socio-technological factors and changing urban spaces. *Space and Culture, 25*(4), 553–564. https://doi.org/10.1177/1206331220910675
Hennig-Thurau, T., Aliman, D. N., Herting, A. M., Cziehso, G. P., Linder, M., & Kübler, R. V. (2022). Social interactions in the metaverse: Framework, initial evidence, and research roadmap. *Journal of the Academy of Marketing Science, 51*, 889. https://doi.org/10.1007/s11747-022-00908-0
Herrman, J., & Browning, K. (2021, July 10). Are We in the Metaverse Yet? *The New York Times*, A1. Retrieved on September 20, 2022 from https://www.nytimes.com/2021/07/10/style/metaverse-virtual-worlds.html
Hwang, I., Shim, H., & Lee, W. J. (2022). Do an organization's digital transformation and employees' digital competence catalyze the use of telepresence? *Sustainability, 14*(14), 8604. https://doi.org/10.3390/su14148604
IJsselsteijn, W., & Riva, G. (2003). Being there: The experience of presence in mediated environments. In G. Riva, F. Davide, & W. IJsselsteijn (Eds.), *Being there: Concepts, effects and measurements of user presence in synthetic environments* (pp. 3–16). IOS Press.
Kang, G., Koo, J., & Kim, Y.-G. (2023). Security and privacy requirements for the metaverse: A metaverse applications perspective. *IEEE Communications Magazine, 62*, 1–7. https://doi.org/10.1109/MCOM.014.2200620
Kaur, P., Dhir, A., Chen, S., Malibari, A., & Almotairi, M. (2020). Why do people purchase virtual goods? A uses and gratification (U&G) theory perspective. *Telematics and Informatics, 53*, 10–21. https://doi.org/10.1016/j.tele.2020.101376

Kim, Y., & Lee, H. (2022). Falling in love with virtual reality art: A new perspective on 3D immersive virtual reality for future sustaining art consumption. *International Journal of Human–Computer Interaction, 38*(4), 371–382. https://doi.org/10.1080/10447318.2021.1944534
Ko, S. H., & Rogers, J. (2021). Functional materials and devices for XR (VR/AR/MR) applications. *Advanced Functional Materials, 31*(39), 2106546. https://doi.org/10.1002/adfm.202106546
Lampe, C., Ellison, N., & Steinfield, C. (2006). A Face(book) in the crowd: Social searching vs. social browsing. In *CSCW'06: Proceedings of the 20th anniversary conference on computer supported cooperative work* (pp. 167–170). https://doi.org/10.1145/1180875.1180901
Lee, H., & Cho, C.-H. (2020). Uses and gratifications of smart speakers: Modelling the effectiveness of smart speaker advertising. *International Journal of Advertising, 39*(7), 1150–1171. https://doi.org/10.1080/02650487.2020.1765657
Lee, K. C., & Chung, N. (2008). Empirical analysis of consumer reaction to the virtual reality shopping mall. *Computers in Human Behavior, 24*(1), 88–104. https://doi.org/10.1016/j.chb.2007.01.018
Liu, Y., & Shrum, L. J. (2002). What is interactivity and is it always such a good thing? Implications of definition, person and situation for the influence of interactivity on advertising effectiveness. *Journal of Advertising, 31*(4), 53–64. https://doi.org/10.1080/00913367.2002.10673685
Maloney, D. (2021). *A youthful metaverse: Towards designing safe, equitable, and emotionally fulfilling social virtual reality spaces for younger users*. Clemson University.
Manis, K. T., & Choi, D. (2019). The virtual reality hardware acceptance model (VR-HAM): Extending and individuating the technology acceptance model (TAM) for virtual reality hardware. *Journal of Business Research, 100*, 503–513. https://doi.org/10.1016/j.jbusres.2018.10.021
Marr, B. (2021). *Extended reality in practice: 100+ amazing ways virtual, augmented and mixed reality are changing business and society*. Wiley.
Matusitz, J. (2021). *Communication in Global Jihad*. Routledge.
Miao, F., Kozlenkova, I. V., Wang, H., Xie, T., & Palmatier, R. W. (2022). An emerging theory of avatar marketing. *Journal of Marketing, 86*(1), 67–90. https://doi.org/10.1177/0022242921996646
Muñoz-Saavedra, L., Miró-Amarante, L., & Domínguez-Morales, M. (2020). Augmented and virtual reality evolution and future tendency. *Applied Sciences, 10*(1), 322. https://doi.org/10.3390/app10010322
Mystakidis, S. (2022). Metaverse. *Scholarly Community Encyclopedia, 2*(1), 486–497. https://doi.org/10.3390/encyclopedia2010031
O'Brien, M., & Chan, K. (2021, October 28). EXPLAINER: What is the metaverse and how will it work? *ABC News*. Retrieved on August 24, 2022 from https://abcnews.go.com/Business/wireStory/explainer-metaverse-work-80842516
O'Riordan, S., Feller, J., & Nagle, T. (2016). A categorisation framework for a feature-level analysis of social network sites. *Journal of Decision Systems, 25*(3), 244–262. https://doi.org/10.1080/12460125.2016.1187548
Papacharissi, Z. (2010). *A private sphere: Democracy in a digital age*. Polity Press.
Park, S.-M., & Kim, Y.-G. (2022). A metaverse: Taxonomy, components, applications, and open challenges. *IEEE, 10*, 4209–4251. https://doi.org/10.1109/ACCESS.2021.3140175
Pesce, M. (1995). *VRML browsing and building cyberspace*. New Riders Publishing.
Pooyandeh, M., Han, K.-J., & Sohn, I. (2022). Cybersecurity in the AI-based metaverse: A survey. *Applied Sciences, 12*(24), 12993. https://doi.org/10.3390/app122412993
Qamar, S., Anwar, Z., & Afzal, M. (2023). A systematic threat analysis and defense strategies for the metaverse and extended reality systems. *Computers & Security, 128*(2), 103127. https://doi.org/10.1016/j.cose.2023.103127
Ravenscraft, E. (2022, April 25). What is the Metaverse, exactly? *Wired*. Retrieved on May 19, 2023 from https://www.wired.com/story/what-is-the-metaverse/
Real de Asúa, E. M., Otter, V., Tsukuda, T., & Vivenot, B. (2022). *The metaverse challenges and regulatory issues*. Retrieved on November 1, 2022 from https://www.sciencespo.fr/public/sites/sciencespo.fr.public/files/Metaverse-Group-report-final-draft-June-12-1.pdf

Riar, M., Korbel, J. J., Xi, N., Zarnekow, R., & Hamari, J. (2021). The use of augmented reality in retail: A review of literature. *Proceedings of the Annual Hawaii International Conference on System Sciences, 54*, 638–647. http://hdl.handle.net/10125/70689

Ruan, J., & Xie, D. (2021). Networked VR: State of the art, solutions, and challenges. *Electronics, 10*(2), 166. https://doi.org/10.3390/electronics10020166

Schegloff, E. A. (2002). Reflections on research on telephone conversations: Issues of cross-cultural scope and scholarly exchange, interactional import and consequences. In K. K. Luke & T. Pavlidou (Eds.), *Telephone calls: Unity and diversity in conversational structure across languages and cultures* (pp. 249–282). John Benjamins.

Sivasankar, G. A. (2022). Study of blockchain technology, AI and digital networking in metaverse. *IRE Journals, 5*(8), 110–115.

Statista. (2021). Roblox Gamers Users Distribution Worldwide September 2020, by Age. Retrieved on October 4, 2022 from https://www.statista.com/statistics/1190869/roblox-games-users-global-distribution-age/

Stephenson, N. (1992). *Snow crash*. Bantam Books.

Suh, A., & Prophet, J. (2018). The state of immersive technology research: A literature analysis. *Computers in Human Behavior, 86*, 77–90. https://doi.org/10.1016/j.chb.2018.04.019

Sykownik, P., Maloney, D., Guo, F., & Masuch, M. (2022). Something personal from the metaverse: Goals, topics, and contextual factors of self-disclosure in commercial social VR. *Proceedings of the 2022 CHI Conference on Human Factors in Computing Systems, 632*, 1–17. https://doi.org/10.1145/3491102.3502008

Tang, F., Chen, X., Zhao, M., & Kato, N. (2022). The roadmap of communication and networking in 6G for the metaverse. *IEEE Wireless Communications, 30*(4), 72–81. https://doi.org/10.1109/MWC.019.2100721

Uspenski, I., & Guga, J. (2022). Embodying metaverse as artificial life: At the intersection of media and 4E cognition theories. *Philosophy and Society, 33*(2), 326–345. https://doi.org/10.2298/FID2202326U

Vita-More, N. (2010). Epoch of plasticity: The metaverse as a vehicle for cognitive enhancement. *Virtual Creativity, 1*(1), 69–80. https://doi.org/10.1386/mvcr.1.1.69_1

Walther, J. B., Loh, T., & Granka, L. (2005). Let me count the ways: The interchange of verbal and nonverbal cues in computer-mediated and face-to-face affinity. *Journal of Language and Social Psychology, 24*(1), 36–65. https://doi.org/10.1177/0261927X04273036

Xi, N., & Hamari, J. (2021). Shopping in virtual reality: A literature review and future agenda. *Journal of Business Research, 134*(1), 37–58. https://doi.org/10.1016/j.jbusres.2021.04.075

Yim, M. Y.-C., Chu, S.-C., & Sauer, P. L. (2017). Is augmented reality technology an effective tool for E-commerce? An interactivity and vividness perspective. *Journal of Interactive Marketing, 39*, 89–103. https://doi.org/10.1016/j.intmar.2017.04.001

Zhao, Y., Jiang, J., Chen, Y., Liu, R., Yang, Y., Xue, X., & Chen, S. (2022). Metaverse: Perspectives from graphics, interactions and visualization. *Visual Informatics, 6*(1), 56–67. https://doi.org/10.1016/j.visinf.2022.03.002

Chapter 3
Communication in the Metaverse: General Perspectives

The concept of communication is of utmost importance in this book because the decentralized nature of the metaverse—and what it will become in the future—is weakening the power of mass communication and will fundamentally alter other forms of communication—particularly interpersonal, intercultural, and health communication (which are described in this chapter). On the whole, communication refers to the exchange of ideas and meanings through feelings and behaviors from one individual to another. Communication is often compelling and intends to receive a desirable response to what is being imparted. Communication is not a one-way street, but based on two or multiple ways. The resulting purpose is that information transmitted must be correctly interpreted by the receiver. Hence, there are senders and receivers. Communication, then, is the process of sending and receiving messages through verbal or nonverbal means, including oral methods, written methods, signs, symbols, and so forth (Kumari & Gangwar, 2018).

Communication is our deepest connection to humankind. Because it is a process that consists of transferring meaning (deliberately or accidentally), when one individual does or says something, he or she creates symbolic behavior, while the recipients discern what was done or said and ascribe meaning to it. After attributing meaning to words or actions, back-and-forth communication can take place (Gamble & Gamble, 2013). Humans are involved in one or multiple communities and societies; they form them and, in turn, are shaped by them. The interaction with others is an existentially important component of their day-to-day living. The human brain enables us to think critically and creatively in manners that require no preprogramming (Raich et al., 2021). The second half of this chapter addresses the notion of social presence in the metaverse. Social presence can be defined as the subjective experience of communicating in one place or environment, even though a person is physically present in another. Communication in the metaverse turns the concept of presence into a multifaceted perception emerging through a mesh of multisensory information and cognitive processes (Witmer & Singer, 1998).

J. Matusitz, J. Dacas, *Communication in the Metaverse*, Signals and Communication Technology, https://doi.org/10.1007/978-3-031-63485-7_3

Interpersonal Communication

Interpersonal communication refers to verbal and nonverbal interactions between two or more individuals, whether it is in a face-to-face or one-on-one situation. The main features of interpersonal communication include: (1) *Source-receiver*. Each has both the task of producing and transmitting information, and the task of receiving and interpreting information; (2) *Encoding–decoding*, which denotes the act of creating and understanding a message; (3) *Message*, which includes both verbal and nonverbal language (e.g., gestures, hearing, vision, touch, smell, and taste); (4) *Channels*. Communication generally occurs simultaneously through different channels; (5) *Noise*, particularly physical, physiological, psychological, and semantic noise; (6) *Context*, which alludes to physical context, temporal context, psychosocial context, and cultural context; (7) *Ethics*; and (8) *Competence*. All eight features attest to our abilities to communicate interpersonally (Matusitz, 2013).

What It Means for the Metaverse

A big question looms in the horizon: Will we be able to accomplish elevated interpersonal communication thanks to those eight abilities? It is fair to contend that the ideal concretization of the metaverse is contingent upon the social element of technology innovation and the participative and humanistic element of technology design. Considering that fundamental human communication in the physical world occurs in various forms, at this stage a key issue is that virtual humans are not able to express the entire spectrum of subtle gestures and mannerisms that typify a real human (e.g., body language and facial expressions). During human interactions, users with rigid postures, whether in fully virtual environments or mixed media, influence communication and, therefore, have an impact on immersion for both the user and the viewer (Bibri, 2022).

The ancient Greek philosopher Aristotle called human beings "social animals." In the face of the many hurdles that prevent them from having enough latitude to interact and socialize, they have found ways to remedy these challenges. This includes the use of social media. Nevertheless, these remain sufficient because they cannot fill the void of being close (Keil & Kreft, 2019). Social networking sites have reconfigured priorities grounded in a simple communicative model: a model for users who post or exchange short-form content. This popular crowd-sourced method of information circulation, in concert with the short-form, high-speed, "disposable" medium of social media, cannot measure up with the metaverse (Caplan et al., 2018) because the former is modeled by the big-tech leaders who recreate narrowcasting in a face-to-face environment. We might look at these technological media in terms of their complementarity: on the one hand, the availability of asymmetric modes of communication; on the other hand, the assurance of a highly communicative metaverse; one that transcends present-day communication barriers (Andrejevic & Volcic, 2020).

An Imaginary Space of Interpersonal Communication

To this day, the metaverse remains an imaginary space of interpersonal communication, where messages are not only disseminated through words but also through pictures, videos, sounds, and feelings. This is a throwback to the classic model of communication described earlier: messages are transmitted by the sender, the medium will also determine the quality and speed of message, and the recipient will have to interpret/understand it. Predominantly, the metaverse is an environment for users to imagine, produce, and share their joint space media. This classic framework of interpersonal communication helps explain how individuals send and receive messages. It is based on the conjecture that interactants can understand messages if both parties agree to the rules of the game (Matusitz, 2022).

Digital platforms in the metaverse have the ability to transform our thoughts and actions by fashioning a new context that allows us to think and act in even more novel ways. Although definitions of the metaverse vary from case to case, the metaverse is foundationally an interactive, immersive, and hyperrealistic world that is shared and virtual. Many like to think that the metaverse is a "place." While this view is not erroneous per se, it is incomplete because it also denotes an era, age, or stage in which immersive digital platforms are becoming the principal manner in which people interact and share digital experiences (Alawaad et al., 2022). How communication in the metaverse can construct a more humane world is a hot debate. Broadly speaking, interpersonal communication can help accomplish this objective by leveraging one-on-one communication styles through verbal communication and body language. Benefiting from these types of interaction is crucial to the long-term aspiration to a triumphant and fully functioning metaverse. Communication through text (as input) is certainly the first step. However, more profound, complex, integrated human behavior input—such as the many forms of nonverbal communication (e.g., facial expressions, body posture, and eye gaze)—not only enhances the metaverse but also buttresses high-quality human–social interactions at large (Wang et al., 2022).

Intercultural Communication

Intercultural communication is a normal occurrence in modern societies across the world. It remains a concept that it is difficult to define because it is not simply about cultural differences, but it is also about national origins, ethnicity, faith, gender, and sexual orientation, among others (García-Jiménez et al., 2017). As Lustig and Koester (2003) state, "intercultural communication occurs when large and important cultural differences create dissimilar interpretations and expectations about how to communicate competently" (p. 51). Intercultural communication is a bilateral course of action beneficial not only to the development of the communicative abilities of nonnative English speakers but also to native English speakers when interacting with other cultures (Matusitz, 2015). It can enhance the expectations and experiences of English- and non-English-speaking interactants (Bao, 2023).

Adopting Multiple Ways of Communicating

During intercultural communication encounters, any attitude too confrontational or too reserved is generally less desirable. By engaging in (1) a process of smooth communication, (2) multiple avenues of sharing points of view, (3) adaptation to the "Other" audience, (4) constant improvement of one's communicative skills, and (5) efforts to understand the diverse personalities from other cultures can interactants genuinely better their understanding of the destination culture. By taking all these five elements into account in the metaverse, we can also take advantage of the absence of the three-dimensional (3D) restrictions of time, space, physiology, and social distancing by allowing users from global cultures to meet their desire for expression on a global digital platform. In this manner, the metaverse will facilitate communication through oral, written, and nonverbal expression that will generate a universal exchange system (Bao, 2023).

This prospect of a global intercultural metaverse is underway; it is already extolled to be an answer to the social and interaction challenges that users have experienced. To this point, many proponents believe that individuals will experience unparalleled virtual social presence through extended reality (XR) technologies. Accordingly, users will have plenty of occasions to interact, play, organize social events, contemplate new dimensions, and collaborate in the virtual environment, such as they did in the physical world. Of equal relevance is the fact that social barriers such as racial bigotry and segregation may be lessened, particularly if users have the ability to project themselves through avatars (cited in Allam et al., 2022, pp. 780–5).

The Effect of Ideology

Audiences perceive data and visual stimuli differently based on their cultural backgrounds. This issue is compounded when people's cultural heritage shapes their ideologies and, in so doing, transform the dominant structures, processes, and/or practices of a space or situation (Arnesson, 2023). Within the context of this book, ideology is broadly defined as a set of beliefs that encompass faiths, political beliefs, and personal convictions. Ideology is a useful concept that helps contextualizing the intercultural communication process in the metaverse. It is inherent to the method by which intercultural relations are created and perpetuated (Bao, 2023). The heart of ideology is situated within a spiral that includes communication and power relations among interactants (Mumby, 1989). It also alludes to wide-ranging social effects that present themselves (to the parties involved) as simply "the way things are." In most cultures, ideologies are inculcated and imposed hegemonically; that is, they frequently get influenced by the confines of social power (Bizzell, 1989). An important corollary is that, to a certain degree, even in the metaverse it can affect intercultural communication among users.

Health Communication

Health communication is understood to mean the use of communication processes to inform choices that influence health (Matusitz & Spear, 2014, 2015). In the metaverse, health communication can take the form of avatar-based communication. For example, 3D avatars of health workers can assist in the collaboration with tools such as digital whiteboards, and they can meet face-to-face without any expensive or complicated conferencing equipment. Healthcare devices, instruments, systems, and procedures can be safely tested via digital twins to identify errors and vulnerabilities before utilizing them in a physical environment (Thomason, 2021). By extension, a metaverse healthcare application could include producing a digital copy of a hospital transaction, such as in-patient flow. In the same vein, such healthcare application could involve applying advanced analytics and probing countless potential scenarios to pinpoint the root cause and run different interventions to determine the best course of action (Rizk, 2021). Here, health communication in the metaverse transcends eHealth—the use of distant communication technologies in healthcare and the use of advanced information and communication technologies (i.e., the internet)—and health information exchange—the sharing of healthcare information electronically across groups or institutions (Matusitz & Breen, 2007a, b).

Enhanced Collaboration

Because the metaverse can improve healthcare collaboration through myriad technologies, learners could be grouped or categorized based on their performances or skills. Put differently, they could form groups with homogeneous levels of accomplishment. For instance, new EdTech platforms such as Studyum—a virtual platform that enables learning through artificial intelligence, facial recognition, and 3D avatars (Pandey et al., 2023)—can create spaces to organize collaborative activities based on the learners' (and teachers') abilities. Studyum could be fully incorporated into healthcare curricula. This type of intracommunity collaboration can be incentivized with other "tokens" and motivators to enhance information exploration and sensemaking, where aspiring or experienced healthcare workers (or scientists) can find new ways to explore data, produce hypotheses, develop solutions, and build knowledge and databases. The metaverse provides a scope with the potential to 3D-model almost anything, where real-world applications can be effectuated via technologies such as digital twins (cited in Thomason, 2021, p. 14).

Enhanced collaboration between health workers also means enhanced medical visits. The metaverse offers a virtual platform where doctors can effortlessly interact with each other without experiencing the usual challenges in the physical world. They can easily deliberate about any treatment process or medication (Mozumder et al., 2022). People with heart problems ought to undergo virtual consultations in lieu of physical visits. Those suffering from depression or mental issues can benefit from the services of a medical expert in this digital environment (even without divulging their identity). Many other types of patients can interact with their physicians in the metaverse, particularly those living in remote areas because it is difficult for them to go to the hospital. These physicians can still perform routine body tests such as testing blood pressure, body temperature, and heartbeat (Mozumder et al., 2022).

AR in Health Communication

There is ample evidence that augmented reality (AR) can generate realistic experiences with genuine emotional, cognitive, social, and behavioral responses (Llobera et al., 2010). The substance in AR can be altered in many ways, including those for ideal health communication, with three main application domains. First, AR allows healthcare workers and consumers to gain from circumstances that they would not in the real world. A particular case would be observing how otherwise unobservable viruses or bacteria proliferate (Vandeweerdt et al., 2022). Second, AR can defeat spatial and temporal barriers by imparting knowledge about the negative effects of unhealthy behaviors that are transiently distant from the present behavior, as would be the case for obesity and smoker's lungs (Jung et al., 2019).

Third, AR enables healthcare workers to experience the world from novel approaches, especially those of vulnerable patients, thereby bringing forth empathy and compassion (Herrera et al., 2018). The first use of AR was conducted by the Johns Hopkins Medical lab in mid-2020. It was for a surgical operation of six vertebrae of a patient's spine (a type of fusion) who was suffering from back pain. During the entire operation, the medical lab relied on see-through eye displays, which was useful in projecting anatomy images analogous to X-ray vision (Locurcio, 2022). Metaverse can boost the productivity and effectiveness of surgical procedures by turning CT scans into a 3D restoration, which assists surgeons in reshaping and relocating the essential parts over the course of important operations (Taheri & Kalnikaite, 2022).

The metaverse can also play a crucial part in the specialty of plastic surgery, where avatars can accurately predict the outcomes of conceivable plastic surgeries.

This is true for both simple and complex surgeries, such as the ones involving the removal of brain tumors and performing difficult spinal surgeries (Elmer et al., 2023). In the specialty of radiology, the metaverse helps radiologists visualize clearer and more compelling images, leading to fewer erroneous disease findings and, as a result, a higher incidence of error-free decision-making. In this vein, radiologists from all over the world can join forces and train each other to increase their skills at using AR medical images for better treatment results. Metaverse dashboards present a great opportunity for medical practitioners to meet with their patients. Conversely, patients can receive extra care by giving doctors updated information in the AR environment, as opposed to other traditional methods of health communication (Garavand & Aslani, 2022).

Social Presence in the Metaverse

Described as an individual's perception of "being there" or being absorbed/engrossed in a medium, presence is a powerful psychological condition for all types of communication mediated in the digital environment. It is also at the center of the difference between experiences in virtual reality (VR) headsets and those in other forms of media territories (Bailenson, 2018). Regarding social interactions between two or more interactants in media environments, Oh et al. (2018) affirm that social presence plays a similarly fundamental part. Whereas physical presence refers to a place's location or geography at a specific time, social presence is more fluid because it denotes an individual's perception of "being (somewhere) together" with others (Biocca et al., 2003). This is why it is affected by the volume and intensity of social cues conveyed by others (Short et al., 1976). Scholars such as Daft and Lengel (1986) have drawn on social presence theories to assert that, as virtual-reality environments are "richer" (i.e., by offering more space and latitude for cues than 2D media), users can transmit not only social cues through text and audio cues but also multifaceted visual and even haptic (touch) sensations (see also Schroeder, 2002). Hence, the metaverse is a facilitator for high degrees of social presence (Wedel et al., 2020), which Smith and Neff (2018) even connect to gatherings in the real world.

The Innate Need for Social Connectedness

We, earthlings, have an intrinsic need for social connectedness. It is believed to shape the internal states of a human being such as the ability to perceive and understand others' ideas, intentions, and behaviors, thereby modifying the decision-making process (Soiné et al., 2021). The concept of social presence is vital for successful VR environments such as online teaching/learning, online shopping, tele-collaborations, gaming, assisted driving, and so forth (Cui et al., 2013). In contrast to telepresence and self-presence, social presence necessitates the actual experience in digital conditions moderated by a copresent entity. Social presence also refers to an individual's state in an environment (not just his or her presence itself)—the state of consciousness in such digital conditions and the sense of being in a place (Slater & Wilbur, 1997).

The minimum degree of social presence happens when users have the sensation that a form, behavior, or emotional experience signals the presence of another intelligence. The degree of social presence is the extent to which a user feels access to the brainpower, wishes, and sensory impressions of another (Biocca, 1997). More descriptions of social presence can be found in other scholars' works (e.g., see Kreijns et al., 2022). Presence, by its very nature, is a complex term to describe or measure. A standard tool for evaluating presence was developed by Lombard and Ditton (1997). Their model listed six dimensions for presence: social richness, realism, transportation, immersion, social actor within medium, and medium as social actor.

Three Categories of Social Presence

Understanding the communicative experience of presence is essential to the design of advanced virtual environments. Over the past few decades, certain scholars (e.g., Kim & Biocca, 1997) have striven to operationalize the concept of presence. Schubert et al. (2001) classifies social presence into three categories: (1) spatial presence, (2) involvement, and (3) experienced realism. Spatial presence is the physiological sensation when being located in a virtual environment, involvement indicates the personal concentration on the virtual environment (and, therefore, feeling of participation), and experienced realism is the individual's subjective experience of the virtual environment. For Honegger et al. (2021), experienced realism is usually formed through auditory stimuli. In their study, they remark that the impact and perception of presence in experiential realism does not only hinge on merely adding higher levels of sensory stimuli, but it also hinges on certain combinations of stimuli (Han et al., 2022).

Unlike spatial presence (the "sense of being there"), social presence symbolizes the idea of "being together" (De Greef & Ijsselsteijn, 2001). Social presence relies on communicative signals such as voice and nonverbal cues (Ijsselsteijn et al., 2000), on undivided and divided interaction (Schultze & Brooks, 2019), and on the type of task (among other factors) (Oh et al., 2018). As opposed to the feeling of merely participating in the virtual environment (spatial presence), the sense of being together "there" (virtual togetherness) presupposed the presence of other people. Virtual togetherness is determined by the sense of being in that virtual space and the communication between the users in that space (Rothe et al., 2021).

Argument for a Positive Impact of Social Presence

A common line of reasoning regarding the positive effect of social presence on interaction outcomes is that users' perception of "togetherness" makes the sharing of arguments, thoughts, and feelings more open and candid, which then leads to many possibilities for interactions. Social presence is similar to basic social relationship concepts, including relational closeness, a perceptual state that deals with the exchange of innermost feelings and thoughts (Hennig-Thurau et al., 2012). Much the same as closeness, social presence allows more intimate exchanges because high degrees of social presence enable interactants "to act out and express their sense of 'closeness' or intimacy" (Baldassar, 2008, p. 261). In line with these contentions, scholars put forward the idea that intimacy represents an aspect or dimension of social presence (Sung & Mayer, 2012). Accordingly, in a metaverse environment, the greater the quality of interaction afforded by the metaverse technology, the more realistic and instant the interaction with avatars in the overall metaverse. To boot, the greater the quality of interaction, the more ways available for interactions with other people's avatars (Zhang et al., 2022).

Immersion

One striking feature of the metaverse is its power to provide immersion (Steffen et al., 2019). Immersion has to do with the technical affordances of the VR hardware and software to give an image/aura of reality to users' senses (Parong et al., 2020). Immersion is important because of its positive effect on the user's presence in a digital world (Peukert et al., 2019). Alternatively, presence refers to the sense of being "there" in an environment and involves a state of consciousness shaped by the experience of surroundings (Steuer, 1992). The higher the levels of presence, they more likely the users behave in ways similar to their behavior in the physical world, collapsing the two realities into one. Reaching a level of presence is not only the fundamental goal of VR but also its defining characteristic (Steffen et al., 2019).

Social presence also denotes an interpersonal relation on the whole, where verbal and nonverbal communication are the centerpiece for the interaction (Oh et al., 2018). It is crucial in a shared virtual interaction with the feeling of togetherness—to not only a group, but also the entire metaverse community because of its immersion. Immersion, then, can be described as the quantifiable technological capacity of a platform to offer an all-encompassing, surrounding, and true-to-life illusion of reality to all interactants involved. The perception of the virtual grows with the ability to be present and engaged in such virtual environment; thus, technologies such as VR headsets seem more intuitive, realistic, and appealing (Slater & Wilbur, 1997).

Social Self-efficacy

Social self-efficacy generally combines people's relationships and self-efficacy. By definition, it indicates individuals' faith in their own abilities to take part in social interactive tasks that are necessary to maintain personal relationships (Smith & Betz, 2000). The social presence that interactants feel in the metaverse markedly moderates the impact of the proportion of fellow users and the amount of time they contribute to the platform. Not surprisingly, the higher the amount of supportive interactions, the more the users' social self-efficacy. Prior studies have stressed the role of social presence in forecasting a variety of communication outcomes such as perceived credibility (Fogg & Tseng, 1999), learning (Oztok et al., 2015), and relationship satisfaction (Richardson & Swan, 2003).

These examples corroborate the value of social presence in establishing the valence of interactions in which users engage in the metaverse. They become galvanized in managing their impression and synchronizing their communication, particularly when the impression they make on others is pertinent to the fulfillment of their primary goals (Leary & Kowalski, 1990). Now that it has been established that social presence stimulates users' need for social approval (Lee & Park, 2014), it is highly probable that the increased sense of social presence that users enjoy in the metaverse makes their relationship goals more realistic and, thereby, makes supportive communication easier (Oh et al., 2023).

Improving Interactions

As the number of supportive interactions in the metaverse is highly correlated to users' perceptions of social self-efficacy, positive experiences of such interactions may increase users' self-confidence in developing social relationships with others in everyday life. Humans improve social self-efficacy by amassing successful social experiences (Gecas, 1989). This is why it is highly probable that sharing various forms of social support with others in the metaverse helps acquire a sense of control in creating positive social relationships, which remains a major source of social self-efficacy (Bandura et al., 1999). Furthermore, by using their avatars, users can enjoy different forms of supportive interactions that involve verbal and nonverbal communication akin to real-world communication (instead of text-based interactions) (Antonijevic, 2008). This live communication experience also improves social skills (Peterson, 2006). These presuppositions add to the current literature on virtual platforms in the metaverse; that is, these platforms are capable of socially benefitting new generations of users—not only by becoming new spaces for virtual social interactions, but also by enhancing users' competence in handling and controlling offline social relationships (Oh et al., 2023).

Predicting Users' Feelings of Loneliness

Social self-efficacy mediates the impact of the positive interactions in which users engage in the metaverse and their sentiments of loneliness. Defying all prior expectations, the number of supportive interactions in the metaverse decrease users' feelings of loneliness in the real world. Positive relationships are significant as a result of social self-efficacy. The relationship between positive interactions and loneliness grows positive over time, indicating that the more users support others in the metaverse, the more they feel loneliness in their offline lives. More recent studies (e.g., see Beehr et al., 2010; Bolger et al., 2000; Vangelisti, 2009) have disclosed the unwished-for outcomes of negative or unmatched supports. Examples include situations in which users become too dependent on support providers, feel embarrassed asking for support, and see themselves as incompetent.

The situations described above presumably intervene in the process of developing self-confidence in social relationships. It is also probable that certain types of supportive interactions in the metaverse are not advantageous to users, particularly youths. Altogether, supportive interactions among users can help aggravate users' feelings of loneliness if they do not ameliorate their social competence in offline relationships. This set of circumstances underscores the necessity to further examine the different types of services, activities, and interactions that can be used in those virtual environments to heighten users' competence in developing and controlling social relationships (Oh et al., 2023).

Case Study 1: Protection Motivation Theory (PMT) to Explain Social Presence in Health Communication

According to Protection Motivation Theory (PMT), a long-established and widely adopted theoretical model for disease prevention and health promotion, productive or beneficial health communication entails improving threat appraisal and coping appraisal. Threat appraisal hinges on two elements: threat severity (i.e., the perceived degree of severity of the threat to health) and vulnerability (i.e., the perceived degree of vulnerability to the threat). Coping appraisal includes response efficacy and self-efficacy. Response efficacy is a person's belief as to whether a suggested coping behavior will efficaciously decrease a health-related threat; self-efficacy refers to the faith that the person has in his or her ability to carry out the recommended behavior (Rogers, 1975).

How Does PMT Apply in the Metaverse?

As opposed to social media platforms accessed through standard media (i.e., computer terminals or smartphones), which have played an important role in health communication since the turn of the twenty-first century (Gunasekeran et al., 2022), immersive health communication in the metaverse has the particular advantages of presence, agency, and embodiment (Makransky & Petersen, 2021). These advantages can be harnessed, for example, by developing immersive environments for patient support groups, professional-hosted health communities, or health awareness conference to boost response efficacy, reduce the health threat, and prompt health-related behavior change more efficiently than traditional media (Riva et al., 2016).

Eliciting Higher Social Presence

Unlike standard text-, picture-, or video-based interventions, VR technologies can strengthen the basic premises of PMT by evoking more social presence and the feeling of being "there" in the digital healthcare environment (Cummings & Bailenson, 2016). To this point, and analogous to the three aforementioned categories of social presence, presence in this context of health communication is comprised of three subcategories: spatial presence (e.g., the illusion of presence between healthcare workers [or patients] in a specific place), self-presence (i.e., the perceived genuineness of our self-representation in such digital healthcare environment), and social presence (i.e., the perceived realness of virtual users) (Lee, 2004). High social presence in the healthcare metaverse makes situations—e.g., experiencing the effects of unhealthy behaviors (Ahn, 2015)—feel "closer" and, thereby, augments threat appraisal (Breves & Schramm, 2021). Experiencing the effects of health behaviors is also key to coping appraisal, especially response efficacy, because it enables the patient to strongly experience the consequences of such behaviors from a first-person perspective (rather than reading about them) (Plechatá et al., 2022).

As human beings are social creatures, and they get largely influenced by others' behaviors, heightened social presence, often leading to heightened social influence (Hershfield et al., 2011), contributes a lot to behavioral change. In health communication, it is common to use avatars to personify health situations and influence our decisions. We can, indeed, be influenced by the presence of virtual others because they can also create social norms and conformity (Ajzen, 1991). An illustration of this was reflected in the use of "health-aware" programmed agents, which consisted of interacting with agents who proceeded to exercising and healthy eating (in hopes that patient users would do the same). Put simply, self-efficacy not only plays a role in embracing new, positive behaviors and reducing knowledge-behavior gaps (Rimal, 2000), bit it can also be improved by our past experiences and by observing others who fulfill desired actions (Bandura, 1977).

Case Study 2: Metaverse Social

With the help of a local US nonprofit, called Metaverse Social, a certain number of seniors have subscribed to the virtual platform to become socially and more physically active, despite sitting in a wheelchair or laying down in bed most of the day. Robert Signore of Cary created a platform named "Thrive Pavillion," which allows seniors to meet, learn, and play in the metaverse. Those who have been part of the virtual experience frequently play a vast array of games supplied by a variety of creators. Examples of games are golf, tennis, and card games such as Uno. Those seniors can organize social activities thoroughly and continuously to decrease distance and location restrictions (especially during the pandemic). In this manner, the seniors' social skills can be bettered, and their feelings of loneliness can be assuaged. On some occasions, they partake in learning sessions about diabetes and other health-related matters hosted by an expert. Every member uses an avatar within his or her virtual space, one that bears resemblance to him- or herself. They can interact with others by using either his or her birth name or some other identification (see https://www.wral.com/isolated-seniors-discover-metaverse-social-platform-as-a-way-to-meet-new-friends-in-a-virtual-world/20740838/).

This case study demonstrates that social participation is a foundational aspect of happiness and mental health, particularly for seniors (Dause & Kirby, 2019). If they fail in their social participation, they will suffer from social isolation, which will bear a negative impact on their health and well-being (Nicholson, 2012). Thus, Metaverse Social fulfills the elderly's need to benefit from social activities in that they can chip away at their negative feelings of loneliness and social isolation, heighten their social participation, and, overall, improve their well-being. During an epidemic, when it is challenging to socialize offline, the elderly can only socialize online. With the growing popularity of 5G, science and technology have matured tremendously. Metaverse Social has integrated all these advantages. Into the bargain, a holographic VR platform can be integrated by combining AR, VR, and XR. To this point, holographic virtual imaging technology can be used to make the restoration of a scene real. In Metaverse Social, social interactions between seniors and their friends are not merely mediated by voice, text, pictures, and videos. They can have avatars to walk in the park and play games with their buddies without concerning themselves about time and space. In a nutshell, they can reap the benefits of an immersive social experience (Wang et al., 2022).

As we have seen, Metaverse Social can fulfill the psychological needs of the elderly. During the home isolation period of the COVID-19 pandemic, they could perform social activities securely through metaverse-related technologies and interactive tools, making them feel valued through their fresh sense of social participation. At the same time, the seniors' mental status (before and after use) was quantitatively evaluated; mental changes were observed in detail. The data revealed that social scenes and functions were improved to accomplish higher therapeutic results (Wang et al., 2022).

Case Study 3: Real-Time Multisensory Social Interactions (RMSIs)

Big-tech companies have shown a move toward more immersive venues such as AR and VR for social interactions that will use our five senses fully, such as in real time. The social character of the metaverse is marked by its ability to welcome real-time multisensory social interactions (RMSIs). RMSIs are interactions among people that occur simultaneously and can integrate multiple senses (e.g., sight, hearing, touch) in the same situation. Because RMSIs have captured the attention of big-tech executives, they have visualized the metaverse as an environment in which RMSIs can exceed the expectations that companies have already offered to consumers so far: 2D platforms such as Zoom, Skype, and Discord (Hennig-Thurau et al., 2022b).

Illusion of Non-mediation

RMSIs in the metaverse afford higher levels of social presence for interactants than those on such 2D platforms. Current VR headsets offer realistic and deep illusions of environments (Wedel et al., 2020). The 360-degree nature of such settings allows individuals to "move around," which adds to the range and intensity of social cues that can be displayed and imparted (Oh et al., 2018). These features build an illusion of non-mediation (Lombard & Ditton, 1997) in which users have a limited awareness of assisting VR technologies (Yadav & Varadarajan, 2005). This is another illustration of the solid feelings of "togetherness" among interactants (Grewal et al., 2020). On the other hand, RMSIs in 2D platforms incur "sensory disadvantages" (Steinhoff et al., 2019, p. 375) that limit the ability of computer-mediated environments to mimic high degrees of social presence.

VR technologies integrate such high degrees of social presence by creating significant performance increments in motion tracking. In this context, motion tracking transmits the user's movements from the real world to the virtual environment. Performance increases are also noticed in processing capacity, screen resolution, battery life, and usability (Bailenson, 2018), which have secured apps such as "room scaling"—i.e., an armchair feature whereby users can move openly and willingly in their physical space while traversing virtual worlds—and high fidelity with respect to settings and avatars' natural bodily expressiveness (Nowak & Biocca, 2003).

Embodied Cognition Theory

Moving more openly and willingly during RMSIs should allow interactants to experience positive interaction outcomes. Embodied cognition theory posits that the human body's interaction with its environment leads to a positive state of mind.

Therefore, an individual's environment can invigorate his or her mind by granting precious cues through interactions with it. Such valuable cues involve all five senses, particularly seeing, hearing, and touching. Embodied cognition is not only about interactions with physical environments, but it is also about digital settings. More opportunities for mobility during RMSIs involve higher usage of bodily movements such as head shaking, nodding, and other visual, auditory, and haptic stimuli. This naturally causes more processing, both cognitive and emotional, and results in a better understanding of one's environment. Owing to the higher degree of interactions with one's surroundings and the impact on processing activities, moving more openly and willingly during RMSIs leads, again, to better interaction outcomes (Hennig-Thurau et al., 2022a).

Vis-à-vis performance outcomes, Oppezzo and Schwartz (2014) have observed more occurrences of embodied cognition for individual users who walk (as opposed to those who sit), because movement is more conducive to the free flow of ideas. The impact of this higher processing is not limited to creative endeavors in RMSIs; it also impacts other aspects of performance, such as the amount of exchange among interactants and their productivity. Mobility is also more likely to contribute to more positive evaluations of RMSIs by stimulating more instances of situational involvement within the interactants (Arts et al., 2011; Richins et al., 1992). As such, the higher number of cues perceived in high interactant mobility result in higher involvement and, thereby, more positive evaluations (Pierro et al., 2006).

References

Ahn, S. J. (2015). Incorporating immersive virtual environments in health promotion campaigns: A construal level theory approach. *Health Communication, 30*(6), 545–556. https://doi.org/10.1080/10410236.2013.869650

Ajzen, I. (1991). The theory of planned behavior. *Organizational Behavior and Human Decision Processes, 50*(2), 179–211. https://doi.org/10.1016/0749-5978(91)90020-T

Alawaad, H. A., Mohamed, E. A. S., & Draz, A. M. M. A. (2022). Metaverse: Imagine a spatial medium for your messages according to classical communication models. *Central European Management Journal, 30*(3), 3447–3458.

Allam, Z., Sharifi, A., Bibri, S. E., Jones, D. S., & Krogstie, J. (2022). The metaverse as a virtual form of smart cities: Opportunities and challenges for environmental, economic, and social sustainability in urban futures. *Smart Cities, 5*(2), 771–801. https://doi.org/10.3390/smartcities5030040

Andrejevic, M., & Volcic, Z. (2020). Virtual empathy. *Communication, Culture and Critique, 13*(3), 295–310. https://doi.org/10.1093/ccc/tcz035

Antonijevic, S. (2008). From text to gesture online: A microethnographic analysis of nonverbal communication in the second life virtual environment. *Information, Communication & Society, 11*(2), 221–238. https://doi.org/10.1080/13691180801937290

Arnesson, J. (2023). Influencers as ideological intermediaries: Promotional politics and authenticity labour in influencer collaborations. *Media, Culture & Society, 45*(3), 528. https://doi.org/10.1177/01634437221117505

Arts, J. W. C., Frambach, R. T., & Bijmolt, T. H. A. (2011). Generalizations on consumer innovation adoption: A meta-analysis on drivers of intention and behavior. *International Journal of Research in Marketing, 28*(2), 134–144. https://doi.org/10.1016/j.ijresmar.2010.11.002

Bailenson, J. N. (2018). *Experience on demand: What virtual reality is, how it works, and what it can do*. W. W. Norton.

Baldassar, L. (2008). Missing kin and longing to be together: Emotions and the construction of co-presence in transnational relationships. *Journal of Intercultural Studies, 29*(3), 247–266. https://doi.org/10.1080/07256860802169196

Bandura, A. (1977). Self-efficacy: Toward a unifying theory of behavioral change. *Psychological Review, 84*(2), 191–215. https://doi.org/10.1037/0033-295X.84.2.191

Bandura, A., Freeman, W. H., & Lightsey, R. (1999). Self-efficacy: The exercise of control. *Journal of Cognitive Psychotherapy, 13*(2), 10–21. https://doi.org/10.1891/0889-8391.13.2.158

Bao, J. (2023). Cultivation of the intercultural communication ability of English learners in the epidemic era from the perspective of the metaverse—taking China as an example. *SHS Web of Conferences, 157*, 03008. https://doi.org/10.1051/shsconf/202315703008

Beehr, T., Bowling, N., & Bennett, M. (2010). Occupational stress and failures of social support: When helping hurts. *Journal of Occupational Health Psychology, 15*(1), 45–59. https://doi.org/10.1037/a0018234

Bibri, S. E. (2022). The social shaping of the metaverse as an alternative to the imaginaries of data-driven smart cities: A study in science, technology, and society. *Smart Cities, 5*(3), 832–874. https://doi.org/10.3390/smartcities5030043

Biocca, F. (1997). The cyborg's dilemma: Progressive embodiment in virtual environments. *Journal of Computer-Mediated Communication, 3*(2), JCMC324. https://doi.org/10.1111/j.1083-6101.1997.tb00070.x

Biocca, F., Harms, C., & Burgoon, J. K. (2003). Toward a more robust theory and measure of social presence: Review and suggested criteria. *Presence: Teleoperators and Virtual Environments, 12*(5), 456–480. https://doi.org/10.1162/105474603322761270

Bizzell, P. (1989). Reviewed work(s): The social construction of written communication by Bennett A. Rafoth and Donald L. Rubin. *College Composition and Communication, 40*(4), 483–486.

Bolger, N., Zuckerman, A., & Kessler, R. (2000). Invisible support and adjustment to stress. *Journal of Personality and Social Psychology, 79*(6), 953–961. https://doi.org/10.1037/0022-3514.79.6.953

Breves, P., & Schramm, H. (2021). Bridging psychological distance: The impact of immersive media on distant and proximal environmental issues. *Computers in Human Behavior, 115*(1), 106606. https://doi.org/10.1016/j.chb.2020.106606

Caplan, R., Hanson, L., & Donovan, J. (2018, February 21). Dead reckoning: Navigating content moderation after 'fake news. *Data & Society*. Retrieved on September 20, 2022 from https://datasociety.net/output/dead-reckoning/

Cui, G., Lockee, B., & Meng, C. (2013). Building modern online social presence: A review of social presence theory and its instructional design implications for future trends. *Education and Information Technologies, 18*, 661–685. https://doi.org/10.1007/s10639-012-9192-1

Cummings, J. J., & Bailenson, J. N. (2016). How immersive is enough? A meta-analysis of the effect of immersive technology on user presence. *Media Psychology, 19*(2), 272–309. https://doi.org/10.1080/15213269.2015.1015740

Daft, R. L., & Lengel, R. H. (1986). Organizational information requirements, media richness and structural design. *Management Science, 32*(5), 554–571. https://doi.org/10.1287/mnsc.32.5.554

Dause, T. J., & Kirby, E. D. (2019). Aging gracefully: Social engagement joins exercise and enrichment as a key lifestyle factor in resistance to age-related cognitive decline. *Neural Regeneration Research, 14*(1), 39–42. https://doi.org/10.4103/1673-5374.243698

De Greef, P., & Ijsselsteijn, W. A. (2001). Social presence in a home tele-application. *Cyberpsychology & Behavior, 4*(2), 307–315. https://doi.org/10.1089/109493101300117974

Elmer, N. A., Veeramani, A., Hassell, N., Shiah, E., Manstein, S., Comer, C., Bustos, V., & Lin, S. J. (2023). Impact and implementation of plastic surgery interest groups: National survey of plastic surgery interest group leadership. *Plastic Surgery, 32*(2), 329–338. https://doi.org/10.1177/22925503221101955

Fogg, B. J., & Tseng, H. (1999). The elements of computer credibility. In *Proceedings of the SIGCHI Conference on Human Factors in Computing Systems* (pp. 80–87). https://doi.org/10.1145/302979.303001

Gamble, T. K., & Gamble, M. W. (2013). *Interpersonal communication: Building connection together.* Sage.

Garavand, A., & Aslani, N. (2022). Metaverse phenomenon and its impact on health: A scoping review. *Informatics in Medicine Unlocked, 32*, 101029. https://doi.org/10.1016/j.imu.2022.101029

García-Jiménez, L., Rodrigo-Alsina, M., & Pineda, A. (2017). The social construction of intercultural communication: A Delphi study. *Journal of Intercultural Studies, 38*(2), 22844. https://doi.org/10.1080/07256868.2017.1289903

Gecas, V. (1989). The social psychology of self-efficacy. *Annual Review of Sociology, 15*(1), 291–316. https://doi.org/10.1146/annurev.so.15.080189.001451

Grewal, D., Noble, S. M., Roggeveen, A. L., & Nordfalt, J. (2020). The future of in-store technology. *Journal of the Academy of Marketing Science, 48*(1), 96–113. https://doi.org/10.1007/s11747-019-00697-z

Gunasekeran, D. V., Chew, A., Chandrasekar, E. K., Rajendram, P., Kandarpa, V., Rajendram, M., Chia, A., Smith, H., & Leong, C. K. (2022). The impact and applications of social media platforms for public health responses before and during the COVID-19 pandemic: Systematic literature review. *Journal of Medical Internet Research, 24*(4), e33680. https://doi.org/10.2196/33680

Han, D.-I. D., Bergs, Y., & Moorhouse, N. (2022). Virtual reality consumer experience escapes: Preparing for the metaverse. *Virtual Reality, 26*(4), 1–16. https://doi.org/10.1007/s10055-022-00641-7

Hennig-Thurau, T., Marchand, A., & Marx, P. (2012). Can automated group recommender systems help consumers make better choices? *Journal of Marketing, 76*(5), 89–109. https://doi.org/10.1509/jm.10.0537

Hennig-Thurau, T., Aliman, D. N., Herting, A. M., Cziehso, G. P., Kübler, R. V., & Linder, M. (2022a). *The value of real-time multisensory social interactions in the virtual-reality metaverse: Framework, empirical probes, and research roadmap.* Münster.

Hennig-Thurau, T., Aliman, D. N., Herting, A. M., Cziehso, G. P., Linder, M., & Kübler, R. V. (2022b) "Social interactions in the metaverse: Framework, initial evidence, and research roadmap," *Journal of the Academy of Marketing Science*. https://doi.org/10.1007/s11747-022-00908-0

Herrera, F., Bailenson, J., Weisz, E., Ogle, E., & Zaki, J. (2018). Building long-term empathy: A large-scale comparison of traditional and virtual reality perspective-taking. *PLoS One, 13*(10), e0204494. https://doi.org/10.1371/journal.pone.0204494

Hershfield, H., Goldstein, D., Sharpe, W., Fox, J., Yeykelis, L., Carstensen, L., & Bailenson, J. (2011). Increasing saving behavior through age-progressed renderings of the future self. *Journal of Marketing Research, 48*, 23–37. https://doi.org/10.1509/jmkr.48.SPL.S23

Honegger, F., Feng, Y., & Rauterberg, M. (2021). Multimodality for passive experience: Effects of visual, auditory, vibration and draught stimuli on sense of presence. *Journal of Universal Computer Science, 27*(6), 582–608. https://doi.org/10.3897/jucs.68384

Ijsselsteijn, W. A., de Ridder, H., Freeman, J., & Avons, S. E. (2000). Presence: Concept, determinants and measurement. *Proceedings of SPIE—The International Society for Optical Engineering*, 3959. https://doi.org/10.1117/12.387188

Jung, S., Lee, J., Biocca, F., & Kim, J. W. (2019). Augmented reality in the health domain: Projecting spatial augmented reality visualizations on a perceiver's body for health communication effects. *Cyberpsychology, Behavior, and Social Networking, 22*(2), 142–150. https://doi.org/10.1089/cyber.2018.0028

Keil, G., & Kreft, N. (2019). *Aristotle's anthropology*. Cambridge University Press.

Kim, T., & Biocca, F. (1997). Telepresence via Television: Two dimensions of telepresence may have different connections to memory and persuasion. *Journal of Computer-Mediated Communication, 3*(2). https://doi.org/10.1111/j.1083-6101.1997.tb00073.x

Kreijns, K., Kate, X., & Weidlich, J. (2022). Social presence: Conceptualization and measurement. *Educational Psychology Review, 34*(1), 139–170. https://doi.org/10.1007/s10648-021-09623-8

Kumari, R., & Gangwar, R. (2018). Use of expression based digital pictograms in interpersonal communication: A study on social media and social apps. *International Journal of Innovative Knowledge Concepts, 6*(11), 167–179.

Leary, M., & Kowalski, R. (1990). Impression management: A literature review and two-component model. *Psychological Bulletin, 107*(1), 34–47. https://doi.org/10.1037/0033-2909.107.1.34

Lee, E.-J. (2004). Effects of visual representation on social influence in computer-mediated communication: Experimental tests of the social identity model of deindividuation effects. *Human Communication Research, 30*(2), 234–259. https://doi.org/10.1111/j.1468-2958.2004.tb00732.x

Lee, E.-J., & Park, J. (2014). Enhancing virtual presence in E-tail: Dynamics of cue multiplicity. *International Journal of Electronic Commerce, 18*(4), 117–146. https://doi.org/10.2753/JEC1086-4415180405

Llobera, J., Spanlang, B., Ruffini, G., & Slater, M. (2010). Proxemics with multiple dynamic characters in an immersive virtual environment. *ACM Transactions on Applied Perception, 8*(1), 1–12. https://doi.org/10.1145/1857893.1857896

Locurcio, L. L. (2022). Dental education in the metaverse. *British Dental Journal, 232*, 191. https://doi.org/10.1038/s41415-022-3990-7

Lombard, M., & Ditton, T. (1997). At the heart of it all: The concept of presence. *Journal of Computer-Mediated Communication, 3*(2), 1–43. https://doi.org/10.1111/j.1083-6101.1997.tb00072.x

Lustig, M. W., & Koester, J. (2003). *Intercultural competence: Interpersonal communication across cultures*. Allyn & Bacon.

Makransky, G., & Petersen, G. B. (2021). The cognitive affective model of immersive learning (CAMIL): A theoretical research-based model of learning in immersive virtual reality. *Educational Psychology Review, 33*(1), 937–958. https://doi.org/10.1007/s10648-020-09586-2

Matusitz, J. (2013). *Terrorism & communication: A critical introduction*. Sage.

Matusitz, J. (2015). The acculturative experience of French students in a southwestern university apartment complex in the United States. *Journal of Human Behavior in the Social Environment, 25*(3), 261–274. https://doi.org/10.1080/10911359.2015.1005517

Matusitz, J. (2022). *Fundamentals of public communication campaigns*. Wiley.

Matusitz, J., & Breen, G.-M. (2007a). E-health: A new kind of telemedicine. *Social Work in Public Health, 23*(1), 95–113. https://doi.org/10.1300/J523v23n01_06

Matusitz, J., & Breen, G.-M. (2007b). Telemedicine: Its effects on health communication. *Health Communication, 21*(1), 73–83. https://doi.org/10.1080/10410230701283439

Matusitz, J., & Spear, J. (2014). Effective doctor–patient communication: An updated examination. *Social Work in Public Health, 29*(3), 252–266. https://doi.org/10.1080/19371918.2013.776416

Matusitz, J., & Spear, J. (2015). Doctor–patient communication styles: A comparison between the United States and three Asian countries. *Journal of Human Behavior in the Social Environment, 25*(8), 871–884. https://doi.org/10.1080/10911359.2015.1035148

Mozumder, A. I., Sheeraz, M. M., Athar, A., Aich, S., & Kim, H. C. (2022). Overview: Technology roadmap of the future trend of metaverse based on IoT, Blockchain, AI technique, and medical domain metaverse activity. In *International Conference on Advanced Communications Technology (ICACT)*. https://doi.org/10.23919/ICACT53585.2022.9728808

Mumby, D. K. (1989). Ideology & the social construction of meaning: A communication perspective. *Communication Quarterly, 37*(4), 291–304. https://doi.org/10.1080/01463378909385551

Nicholson, N. R. (2012). A review of social isolation: An important but underassessed condition in older adults. *Journal of Primary Prevention, 33*(2), 137–152. https://doi.org/10.1007/s10935-012-0271-2

Nowak, K. L., & Biocca, F. (2003). The effect of the agency and anthropomorphism on users' sense of telepresence, copresence, and social presence in virtual environments. *Presence: Teleoperators and Virtual Environments, 12*(5), 481–494. https://doi.org/10.1162/105474603322761289

Oh, C. S., Bailenson, J. N., & Welch, G. F. (2018). A Systematic Review of Social Presence: Definition, Antecedents, and Implications. *Frontiers in Robotics and AI, 5*, 114. https://doi.org/10.3389/frobt.2018.00114

Oh, H. J., Kim, J., Chang, J., Park, N., & Lee, S. (2023). Social benefits of living in the metaverse: The relationships among social presence, supportive interaction, social self-efficacy, and feelings of loneliness. *Computers in Human Behavior, 139*(6), 107498. https://doi.org/10.1016/j.chb.2022.107498

Oppezzo, M., & Schwartz, D. L. (2014). Give your ideas some legs: The positive effect of walking on creative thinking. *Journal of Experimental Psychology: Learning, Memory, and Cognition, 40*(4), 1142–1152. https://doi.org/10.1037/a0036577

Oztok, M., Zingaro, D., Makos, A., Brett, C., & Hewitt, J. (2015). Capitalizing on social presence: The relationship between social capital and social presence. *The Internet and Higher Education, 26*(1), 9–24. https://doi.org/10.1016/j.iheduc.2015.04.002

Pandey, A., Chirputkar, A., & Ashok, P. (2023). Metaverse: An innovative model for healthcare domain. *International Conference on Innovative Data Communication Technologies and Application*, 22963529. https://doi.org/10.1109/ICIDCA56705.2023.10099764

Parong, J., Pollard, K. A., Files, B. T., Oiknine, A. H., Sinatra, A. M., Jason, D., Moss, A. P., & Khooshabeh, P. (2020). The mediating role of presence differs across types of spatial learning in immersive technologies. *Computers in Human Behavior, 107*, 106290. https://doi.org/10.1016/j.chb.2020.106290

Peterson, M. (2006). Learner interaction management in an avatar and chat-based virtual world. *Computer Assisted Language Learning, 19*(1), 79–103. https://doi.org/10.1080/09588220600804087

Peukert, C., Pfeiffer, J., Meißner, M., Pfeiffer, T., & Weinhardt, C. (2019). Shopping in virtual reality stores: The influence of immersion on system adoption. *Journal of Management Information Systems, 36*(3), 755–788. https://doi.org/10.1080/07421222.2019.1628889

Pierro, A., Kruglanski, A. W., & Tory Higgins, E. (2006). Progress takes work: Effects of the locomotion dimension on job involvement, effort investment, and task performance in organizations. *Journal of Applied Social Psychology, 36*(7), 1723–1743. https://doi.org/10.1111/j.0021-9029.2006.00078.x

Plechatá, A., Makransky, G., & Böhm, R. (2022). Can extended reality in the metaverse revolutionise health communication? *NPJ Digital Medicine, 5*(1), 132. https://doi.org/10.1038/s41746-022-00682-x

Raich, M., Dolan, S. L., Ulrich, D., & Cisullo, C. (2021). Human uniqueness at the Dawn of intelligent machines. *The European Business Review, 3*, 39–59.

Richardson, J., & Swan, K. (2003). Examining social presence in online courses in relation to students' perceived learning and satisfaction. *Journal of Asynchronous Learning Networks, 7*(1), 68–88. https://doi.org/10.24059/olj.v7i1.1864

Richins, M. L., Bloch, P. H., & McQuarrie, E. F. (1992). How enduring and situational involvement combine to create involvement responses. *Journal of Consumer Psychology, 1*(2), 143–153. https://doi.org/10.1016/S1057-7408(08)80054-X

Rimal, R. N. (2000). Closing the knowledge-behavior gap in health promotion: The mediating role of self-efficacy. *Health Communication, 12*(3), 219–237. https://doi.org/10.1207/S15327027HC1203_01

Riva, G., Baños, R. M., Botella, C., Mantovani, F., & Gaggioli, A. (2016). Transforming experience: The potential of augmented reality and virtual reality for enhancing personal and clinical change. *Frontiers in Psychiatry, 7*, 164. https://doi.org/10.3389/fpsyt.2016.00164

Rizk, S. H. (2021). Ethical and regulatory challenges of emerging health technologies. In I. Vasiliu-Feltes & J. Thomason (Eds.), *Applied ethics in a digital world* (pp. 84–100). IGI Global.

Rogers, R. W. (1975). A protection motivation theory of fear appeals and attitude change1. *The Journal of Psychology, 91*(1), 93–114. https://doi.org/10.1080/00223980.1975.9915803

Rothe, S., Schmidt, A., Montagud, M., Buschek, D., & Hußmann, H. (2021). Social viewing in cinematic virtual reality: A design space for social movie applications. *Virtual Reality, 25*(3), 613–630. https://doi.org/10.1007/s10055-020-00472-4

Schroeder, R. (2002). Social interaction in virtual environments: Key issues, common themes, and a framework for research. In R. Schroeder (Ed.), *The Social Life of Avatars* (pp. 1–18). Springer.

Schubert, T., Friedmann, F., & Regenbrecht, H. (2001). The experience of presence: Factor analytic insights. *Presence Teleoperators & Virtual Environments, 10*(3), 266–281. https://doi.org/10.1162/105474601300343603

Schultze, U., & Brooks, J. A. M. (2019). An interactional view of social presence: Making the virtual other 'real'. *Information Systems Journal, 29*(3), 707–737. https://doi.org/10.1111/isj.12230

Short, J., Williams, E., & Christie, B. (1976). *The social psychology of telecommunications*. Wiley.

Slater, M., & Wilbur, S. (1997). A framework for immersive virtual environments (FIVE): Speculations on the role of presence in virtual environments. *Presence: Teleoperators & Virtual Environments, 6*(6), 603–616. https://doi.org/10.1162/pres.1997.6.6.603

Smith, H. M., & Betz, N. E. (2000). Development and validation of a scale of perceived social self-efficacy. *Journal of Career Assessment, 8*(3), 283–301. https://doi.org/10.1177/106907270000800306

Smith, H. J., & Neff, M. (2018). Communication Behavior in Embodied Virtual Reality. *Proceedings of the 2018 CHI Conference on Human Factors in Computing Systems, 289*, 1–12. https://doi.org/10.1145/3173574.3173863

Soiné, A., Flöck, A. N., & Walla, P. (2021). Electroencephalography (EEG) reveals increased frontal activity in social presence. *Brain Sciences, 11*(6), 731–748. https://doi.org/10.3390/brainsci11060731

Steffen, J. H., Gaskin, J. E., Meservy, T. O., Jenkins, J. L., & Wolman, I. (2019). Framework of affordances for virtual reality and augmented reality. *Journal of Management Information Systems, 36*(2), 683–729. https://doi.org/10.1080/07421222.2019.1628877

Steinhoff, L., Arli, D., Weaven, S., & Kozlenkova, I. V. (2019). Online relationship marketing. *Journal of the Academy of Marketing Science, 47*(3), 369–393. https://doi.org/10.1007/s11747-018-0621-6

Steuer, J. (1992). Defining virtual reality: Dimensions determining telepresence. *Journal of Communication, 42*(4), 73–93. https://doi.org/10.1111/j.1460-2466.1992.tb00812.x

Sung, E., & Mayer, R. E. (2012). Five facets of social presence in online distance education. *Computers in Human Behavior, 28*(5), 1738–1747. https://doi.org/10.1016/j.chb.2012.04.014

Taheri, M., & Kalnikaite, D. (2022). A study of how virtual reality and brain computer Interface can manipulate the brain. *International Conference on Software Engineering and Information Management (ICSIM), 5*, 6–10. https://doi.org/10.1145/3520084.3520086

Thomason, J. (2021). MetaHealth: How will the metaverse change health care? *Journal of Metaverse, 1*(1), 13–16.

Vandeweerdt, C., Luong, T., Atchapero, M., Mottelson, A., Holz, C., Makransky, G., & Böhm, R. (2022). Virtual reality reduces COVID-19 vaccine hesitancy in the wild: A randomized trial. *Scientific Reports, 12*, 4593. https://doi.org/10.1038/s41598-022-08120-4

Vangelisti, A. (2009). Challenges in conceptualizing social support. *Journal of Social and Personal Relationships, 26*(1), 39–51. https://doi.org/10.1177/0265407509105520

Wang, A., Gao, Z., Lee, L.-H., Braud, T., & Pan, H. (2022). Decentralized, not dehumanized in the metaverse: Bringing utility to NFTs through multimodal interaction. *International Conference on Multimodal Interaction*, 1–6. https://doi.org/10.1145/3536221.3558176

Wedel, M., Bigné, E., & Zhang, J. (2020). Virtual and augmented reality: Advancing research in consumer marketing. *International Journal of Research in Marketing, 37*(3), 443–465. https://doi.org/10.1016/j.ijresmar.2020.04.004

Witmer, B. G., & Singer, M. J. (1998). Measuring presence in virtual environments: A presence questionnaire. *Presence, 7*(3), 225–240. https://doi.org/10.1162/105474698565686

Yadav, M. S., & Varadarajan, R. (2005). Interactivity in the electronic marketplace: An exposition of the concept and implications for research. *Journal of the Academy of Marketing Science, 33*(4), 585–603. https://doi.org/10.1177/0092070305278487

Zhang, G., Cao, J., Dong, L., & Qi, J. (2022). Popularity of the Metaverse: Embodied Social Presence Theory Perspective. *Frontiers in Psychology, 13*, 997751. https://doi.org/10.3389/fpsyg.2022.997751

Chapter 4
Social Construction and Self-presentation

Drawing upon the constructionist model of Berger and Luckmann (1966), in their work titled *The Social Construction of Reality: A Treatise in the Sociology of Knowledge*, social construction postulates that meaning, ideas about the universe, habits of everyday communication, and even our self-image/identity develop from social consensus and interactions—rather than strict scientific or logical observations. Individuals interact on the principle that their perceptions of the world are shared with others, and this agreed-upon knowledge of reality is then shored up by these interactions. Interactants share identifiable frameworks of meaning and action about communication. Communication entails a process of producing meaning through reality construction (Mumby, 1989). This process is qualified as social because it can develop only through social interaction, whether real or symbolic. Therefore, the social construction of reality is a dialectical activity in which people operate both as the creators and as products of their society (Adoni & Mane, 1984). The end of the chapter addresses self-presentation strategies employed by metaverse inhabitants to construct their personas.

Context, Structure, and Scripting

As we will see, even in the metaverse, communication and related concepts exist in context, not in isolation. Human beings form connections between self and others through communication, often situationally and through an interchange between individual creativity and structural/social constraint (Eisenberg et al., 2007). This frame of reference offers a middle ground between those who envision human potential as mostly unbridled and those who perceive it as influenced by social and economic conditions (Wentworth, 1980). As Craig (2007) continues, "communication, rather than merely a neutral conduit for transmitting independently existing information, is the primary social process through which our meaningful common

J. Matusitz, J. Dacas, *Communication in the Metaverse*, Signals and Communication Technology, https://doi.org/10.1007/978-3-031-63485-7_4

world is constructed" (p. 127). It is important to consider rules or types of daily communication as progressively built structures as well. While it may seem natural to see the social conditions that constrain our activities as immutable, the social constructionist view argues the opposite (Eisenberg, 2008). That is, a number of behavioral patterns get replicated over time and solidify such as water and ice, implying that interactants no longer see them as social creations, but as irrefutable "reality" (Berger & Luckmann, 1966).

An Ongoing Dialectical Process

When discussing the dialectical process, it is critical to embrace the notion of discourse. Discourse is socially constructed and provides understanding into the language employed in various contexts. Social construction rests on the premise that communication shapes reality (Martin Jr., 2017). It establishes that "reality is both revealed and concealed, created and destroyed by our activities" (Pearce, 1995, p. 89). According to Berger and Luckmann (1991), "society is understood in terms of an ongoing dialectical process composed of the three moments of externalization, objectivation and internalization" (p. 149), which also relates to the individual level, by "externalizing his own being in the social world and internalizes it as an objective reality" (Berger & Luckmann, 1991, p. 149).

This framework delineates the way in which humans begin to explore and comprehend their surrounding world, by absorbing the elements that make up the objective reality and ascribing meanings to them so as to integrate their own systems of values: "Internalization, in this general sense is the basis, first, for an understanding of one's fellowmen and, second, for the apprehension of the world as a meaningful and social reality" (Berger & Luckmann, 1991, p. 150). The process of internalization is the method by which people come to understand their own reality and that of others, contributing to the creation of personal identity. Social construction can be considered theoretically related to symbolic interaction (Collett, 2011).

Scripting

Scripting is the making of a set of guidelines that establish the scene, list actors, and outline-expected behavior. Scripts are not always immutable texts that actors are required to follow. Instead, they are interactionally emergent directives for social consciousness and action, guides that are sufficiently meticulous to offer behavioral insights when unanticipated events occur—yet flexible enough to enable improvisation. Although most scripting activity happens before a performance, it can be improvised as participants interact with one another and the audience (Benford & Hunt, 1992). Scripts are constructed upon "frames" that bestow a collective definition to the situation (Snow et al., 1986). Scripting is a social activity that includes all of the different framing activities and alignment methods. It is different from framing because scripts seek to integrate and synchronize movement activity. Whereas framing elicits actable ideas, scripting moves them somewhat closer to enactment. It assigns roles, formulates dialogue, and directs action (Benford & Hunt, 1992).

Carey (1989) summed up two essential conceptions of communication: transmission, in which communication refers to the sharing of messages (communication = interaction) and ritual, a category of scripting. In this case, communication is a symbolic process that gives rise to society, a discourse where understanding and empathy take place. These definitions delineate two consecutive steps in the process of communication because interaction/transmission is the required starting point for the social construction of meanings and the creation of a more significant encounter with the Other—what is also called ritual (García-Jiménez et al., 2017).

Immanuel Kant, the German philosopher, envisioned the purpose of human society as "maximum individuality within maximum community" (cited in Eisenberg, 1984, pp. 227–42). This dialectical process between self and society is considered an existential problem in social theory, mirroring individuals' foundational yearning for agency and belonging. In addition, this dialectical tension between individual agency and social control typifies our day-to-day social interactions, from medical visits to business meetings, and from attendance at football games to political debates (Eisenberg, 2008).

Why It May Still Differ in the Metaverse

A conceptual analysis that applied the principles of the social construction theory examined the success of collaborations within players' playtime and "guilds" (i.e., groups or small federations) in *World of Warcraft*. The results of the study revealed that players' playtime in guilds is productive because it occurs through a shared understanding—i.e., scripting—that rises above the technical forms of the game. This creation of self-identities in virtual games is facilitated by a narrative in which players merge their perceptions in the game with their real-life experiences as players (through their serious participation). Notwithstanding the fact that this guild of gamers appears like a loose, wobbling, physically absent structure, the metaverse still enables a mirroring of real-world gameplay because the participants follow the procedures—i.e., the rules of the ritual (Hu & Liu, 2022).

Nevertheless, although this dialectical process in the metaverse can become a ritual or script for participants as they interact with others on virtual platforms, the experience may still differ from that of the three-dimensional world. If we look at "communication in the metaverse" as an illustration of modern-day discourse, we should also acknowledge that discourse is time- and culture-specific. It can change over time or based on the context, often leading to social transformations (Burr & Dick, 2017). As of today, the process of interaction in the metaverse has not been made identical to traditional symbolic interaction in the real world. Part of the reason is that meanings emerging through interactions depend, to a large degree, on rituals or scripts with which interactants are familiar. Explained differently, in the three-dimensional world, there already exists a deeply entrenched sociocultural environment replete with popular cultural symbols of mass society—even before any interaction takes place. On the other hand, in the metaverse, this context is not necessarily "given" or available, though it may be improvised by the interactants. This prompts another question about the aspect of the metaverse's "openness": When long-established sociocultural symbols are missing on virtual platforms, in what ways and on what basis can the "consensus" of the metaverse society be attained? (Hu & Liu, 2022).

Intersubjectivity and Social Presence

Similar to social interaction, intersubjectivity posits that meaning comes from interactive interpretation by multiple interactants, not merely from the cognition of a single entity. From this vantage point, during human interactions, interpretation of meaning is collectively subjective—i.e., intersubjective (Miranda & Saunders, 2003). Schütz (1967) opposes beliefs in a purely objective reality. He sees information as units of data that confer different meanings to different people, based on their identities, experiences, or sociocultural backgrounds. To this point, the sociocultural background in which information is shared contributes to its attribution of meaning. This is why Schutz introduced the concept of intersubjectivity to define the understanding that springs up from shared human experiences. Harold Garfinkel (1967) also highlighted the relevance of the subjective aspect of meaning. For the famed US sociologist, interpretation is grounded in the situation in which it is embedded; hence, shared meaning is foundational to day-to-day interactions. This shared meaning is not merely intersubjective agreement, but it is also an allusion to communicative events—i.e., things emerging from a particular context.

A parallel can be drawn with the precept of social presence covered in Chap. 3. Intersubjectivity cannot be completely fathomed without looking at it in the context of social presence. Social presence refers to "the degree to which the medium facilitates awareness of the other person and interpersonal relationships during the interaction" (Fulk et al., 1990, p. 118). It is operationalized as "low to high" in most studies of social presence theory (Short et al., 1976). Face-to-face interactions in the metaverse should be integrated within social presence. Dennis and Kinney (1998) contend that full cognizance of the social presence of a medium will increase our understanding of "person-oriented" or social tasks. Social presence theory follows the tenet that interactants' performance improves when media's ability to offer social presence measures up with the social needs of a task (Christie, 1985). While task-oriented activities such as problem-solving might be fulfilled equally well through any medium, person-oriented ones such as conflict resolution generally require the availability of media high in social presence (Miranda & Saunders, 2003).

Inherent to social presence theory is the assumption that the presence of the sender influences the receiver's understanding of the message. Consequently, intersubjective interpretation is a social process inapt to media low in social presence; rather, it requires reciprocity. Media poor in social presence interfere with such reciprocity or interactivity. Even in "synchronous" situations such as virtual meetings, because group members share information concurrently, they concentrate on particular pieces of information asynchronously. As a result, instant reciprocity is a difficult task to accomplish. Low social presence leads to circumstances in which specific comments are ignored because participants cannot address others' urgency or their emotional reactions (Miranda & Saunders, 2003).

Social Construction of Content in the Metaverse

As the construction of social reality also occurs in the metaverse, it goes without saying that the social construction of content is the result of the construction of that reality. It is made possible not only by using language as its primary means but also by relying on other methods such as work collaboration, educational social learning, interactive game play, and user-generated content (UGC). UGC denotes any type of content created by users on digital platforms that becomes an open source (i.e., it is publicly shared) and accessible to other users (Saura & Bennett, 2019).

UGC in the Metaverse

In the metaverse, UGC can be customized to "user-co-created content." In this manner, it could evolve from "production" to "co-creation," building on an extra layer of production and creation community. In a realm of less sophisticated virtual platforms, content creators would still operate separately, as individuals independent of each other, and even attempting to outsmart each other in their creative conduct and content. However, in this metaverse age, with the free support of open source and blockchain, a new order of co-creation, collaboration, and synergy is being concretized. The possibility for everyone to be a creator in the metaverse is not merely germane to the words, pictures, audio, and video with which we are familiar; intrinsically speaking, it is also about the source of information, data, and even the knowledge of the user. UGC in the metaverse offers a new orientation of co-cultural construction and evaluation. Similarly, it smooths the way for the creation of a new framework of distributed decision-making and distributed communication. And it symbolizes the transition from top-down centralized social structures to bottom-up endogenous ones (Hu & Liu, 2022).

Open-Source Codes

The digitalization of daily interactions progressively translates into content of the virtual world that will replicate that of the real world. This has been demonstrated by the surge of digital processes in the physical world, the regulations for point redemption of merchants, incentive reward mechanisms (e.g., such as those for individual lifestyles), and so on. Hundreds of millions of people are already using comparable open-source codes to support their communities in the metaverse. If truth be told, open-source codes will become the benchmark for a growing proportion of co-constructed information and social affairs of digital residents in the metaverse. Open source is essentially the source of all this. In a society nurtured by open-source codes, the existential self-consuming problems of the virtual environment will gradually be resolved by open, creative, and visionary solutions for external advancements (Hu & Liu, 2022).

Through the social construction of content, the metaverse cannot circumvent its connections to and learning from the three-dimensional world. In spite of all the open-source technologies that are appealingly transformative, they would be of little use in the absence of a social consensus or norms in real life, particularly since both self-identity and community formation depend on source codes as an infrastructural technology. Big-tech developers and providers in the real world are behind the source codes and constantly work on being infrastructure-conscious, or else an immeasurable number of copies of user-created codes would be produced and stored in individual users' inventories (for later use). This would also cause many content creators to shut their accounts, close their businesses, or build invisible walls to forbid access (Hu & Liu, 2022).

Cloud Communities in the Metaverse

Cloud communities are cyberspace groups that are provisional, momentary, flexible, malleable, and inexpensive. Members are free to join or to leave. In addition to creating content, they also take part in decision-making (Quadir & Vijayakumar, 2019). The presence of cloud communities in the metaverse alters the conventional structure and definition of virtual communities. According to Moretti and Schlemmer (2012), cloud communities in VR are communities that produce knowledge, allow members to change content, and create spaces for co-creation. Grieve and Heston (2012) describe cloud communities as ephemeral, outsourced, emotionally bound groups of users who can either be people in a real community or fake personas (such as distributed electronic bodies). Having said that, this does not imply that such communities in the metaverse adopt a pattern of organizational structures different from the real world. As discussed earlier, communication and sanctioning apparatuses that stem from real life play an important role in supervising social content and group cohesion in VR. Besides, technology-mediated contexts strengthen social control and furnish new apps and services for members of cloud communities. For instance, they can adopt the traditional rules of the outside world in order to construct, alter, or remove content in the metaverse (Hu & Liu, 2022).

Social Construction of Identity in the Metaverse

Identity is the product of a combination of individual and social constructionist traits that create, for each of us, a specific configuration (Parmentier & Rolland, 2009). This signifies that every human being is identified either by characteristics that display their involvement in social groups (social identity) or by more personal traits that are indicative of their personal identity (Hogg et al., 1995). There is a high correlation between social identity and self-identity in the sense that both play an essential part in an individual's growth through interaction. According to Parmentier and Rolland (2009), it then seems that identity can be (re-)constructed by a method of a dual process. As they explain below, identity is

> (1) a process of emotional closure, a result of others' reactions to mobilized role identities, leading to available role identities being fixed; (2) a process of imaginative opening up, creation of possible selves, fed by the individual's experiences and imagination, leading to the creation of new role identities when interacting with others (p. 50).

Put simply, the manner in which we understand and see ourselves is the result of both our social world (e.g., "a result of others' reactions") and the sociocultural era in which we live (e.g., "the individual's experiences and imagination" and "new role identities").

Self-concept

A person's self-concept shapes not only his or her behavior but also his or her emotional and cognitive processes. It also affects the outcomes in many life domains, such as social integration, self-esteem, sense of achievement, and joyfulness. In actuality, a great part of an individual's experience with or understanding of their self-concept is an emotional reaction to the evaluations made by others (significant others in particular). This is why the connections that we develop with those who are close to us tend to be a product of our self-concept. The notion of a fixed self, formed at an early age by the sociocultural environment, is a self in-relation-to-others; it is elastic and mutable, whether intentionally or unintentionally (Cooley, 1902/1998). Self-concept is intrinsically hierarchical and multidimensional. It is made up of several components, including social self (which comes from the perceptions that we have of the views of others). In a similar fashion, because humans interact in groups, they may develop a multiplicity of selves to harmonize with the number of distinct groups to which they belong (assuming that their identity and opinions are valued by these groups) (Burns et al., 2018; Williams & Page, 1989).

Blurring the Lines Between Physical and Digital Identities

The increasing dominance of the metaverse blurs the lines between physical and virtual identities in unanticipated and potentially unsettling ways. Social networking sites have already adopted measures to associate physical identities (and data) with VR. Thus, the types of identity created in the metaverse's vision of VR form a nexus between both disembodied and embodied interpretations of digital media. By its very nature, the embodiment of the metaverse through virtual platforms blurs the lines between the physical and digital in manners that integrate themselves within the metaverse's wider focus on diverse types of data collection. The virtual self that forms in this environment is not a singular entity in and of itself, but it generates the connective tissue between the lived experience of VR and information obtained (through digital means) about the identity linked to said experience (Saker & Frith, 2019).

The availability and pervasiveness of the metaverse have the power to constantly restructure the constitutive features of identity with respect to articulating, recording, evaluating, and engaging with our bodily selves and bodily others. For that reason, Haraway's (2006) groundbreaking concept of the cyborg—"a cybernetic organism, a hybrid of machine and organism, a creature of social reality as well as a creature of fiction" (p. 117)—illuminates our understanding of the manifold contours of the complex and overlaying relationships between physical and digital identities. This phenomenon portents a reality in which users are not concerned about having "partial" identities or contradictory ones. Setting the specificity of connections aside, the probable effects of metaverse identities are aggravated by the degree to which digital communication and media tools are more and more consolidated into uses that necessitate close contact with the body. These effects can outwardly be observed through an array of ultramodern digital media innovations (Marzaleh et al., 2022).

Interoperability: Making Identities Similar

Interoperability is the ability for a user to switch seamlessly between platforms with his or her own tools and knowledge. This also means that his or her identity as well as virtual assets can be easily crossed over into other virtual platforms. As users circulate through the immense space of virtual worlds of the metaverse, any piece of information (including ads and brand objects) can incessantly follow users through behavioral tracking. This could lead to two problems. First, such behavioral tracking could impact a person's identity by altering his or her self-concept because that person absorbs new information—some of which can be detrimental (and could also confer a bad image to the identity). Second, when a user revisits a metaverse platform and finds that essential pieces of information are no longer there, he or she may experience identity confusion because of that loss of connection (Ahn et al., 2023). These interoperability problems could become so pervasive that they would make identities go through similar metamorphoses.

On the positive side, the concept of concurrency is something to behold. In this context, concurrency implies that a user's identity (or identities) can be shared in real-time with other users across multiple platforms in the metaverse. This aspect of interoperability forges social interactions that lead to a sense of "now," where digital identities get connected in real time with other users' or with objects (Ahn et al., 2023). It should come as no surprise, then, that metaverse identities are more likely to become more similar to each other than the eight billion different ones in the three-dimensional world today. At the same time, assessing the level of similarity between metaverse identities would be so time-consuming and difficult that it would turn into the content of an entire book in and of itself. Having said that, interoperability across VR platforms also makes identities more similar because it enables audiences to access similar mediated content irrespective of the immersive technology they possess (while upholding the same identity across all platforms). This process is analogous to how advertisers organize their campaigns through different communication options in integrated marketing communications (Ahn et al., 2022).

The interoperable possibilities for user identity are paramount because of another attribute of interactions in the metaverse: social presence. Akin to concurrency, large swaths of users can come together, in embodied form, to forge that sense of social presence. Social presence is deeper than concurrency because the former is more of a transactional model of communication that allows dynamic interpersonal interactions that foster relationship building and community formation through shared identity features (Barnlund, 1970). And shared attention too; just like in the real world, when users' identities mesh well, they dedicate more time and energy to the same event, leading to more emotionally charged interactions, better memories, and even greater behavioral change (Boothby et al., 2014). Such social presence can be revived and shared across platforms, and eventually can create a multilayered sense of experience that will augment the value and presence of the metaverse even more. This phenomenon has been felt in the real world, where shared attention intensifies through social presence. In festivals, concerts, stadiums, public squares, and private spaces, people experience world events with others. They do so across all sensory modalities—experiencing similar sights, sounds, tastes, smells, and textures together. This prospect for "attending with" others can grow exponentially with the emergence of metaverse technologies because interoperability allows such social experience and sharing of attention to be felt across many more modalities and platforms (Shteynberg, 2015).

On a last note, according to US psychologist Carl Rogers (1959), the degree to which a person's self-concept aligns with reality is called congruence; naturally, incongruence is when the opposite occurs. More precisely, in the case of congruence, our self-concept aligns with reality well; with incongruence, reality does not correspond to our self-concept. Some detractors, however, believe that the "ideal" self-concept is not necessarily important because it does not add much to the self-concept; rather, it is only the "actual" or "real" self-concept that matters (Waugh, 2000). Within the context of the metaverse, considering that interoperability allows users to move seamlessly between platforms, thereby increasing the likelihood of making virtual identities more similar, the whole notion of communication in the metaverse flies in the face of traditional views of the individual's self-concept.

Aspects of Selves in the Metaverse

To date, research has established many different categories of selves, which include, but are not limited to, actual, ideal, and ought selves (e.g., see Higgins, 1987), personal, social, and looking-glass selves (Brewer & Gardner, 1986; Mead, 1934), possible selves (Markus & Nurius, 1986), and extended selves (Belk, 1988). After the advent of computers and computer-mediated communication, academics have included other selves, such as the extended-digital self (Belk, 2016), and digital, online, or cyber-selves (Robinson, 2007; Zhao, 2005). According to Belk (2013), the virtual environment is ideal for facilitating the creation and experimentation with multiple personas to the users' wishes (called "multiplicity" by Belk). Bearing in mind that these users are unlikely to actually meet in the physical world, the selves are often perceived as more similar because of the manners in which most metaverse platforms are designed. Table 4.1 presents and discusses the selves that are pertinent to the metaverse (Lloyd, 2019).

These selves can be regarded from either a "self-concept" view or from the view of the other individual in the metaverse. As a brief reminder, one's self-concept personifies how a user sees him- or herself. Even though this self can be seen by the outside world, it is expected that it diverges from the self in the real world. One's ideal self exists in a person's dreams and aspirations, or the set of skills, characteristics, and resources which that person seeks to obtain (Markus & Nurius, 1986). In the metaverse, images of the ideal self are much more achievable because they can be fashioned through physical and mental representations (i.e., avatars) (Rusbult et al., 2009). What this table also demonstrates is that assuming any facet of the self in the metaverse does not, in any way, embroil how the self functions in real life. Instead, any digital self is grounded in a connective tissue between our already-existing social media appearances and activities, our colossal volume of digital personal data, and other interactions on virtual platforms. Put differently, the metaverse self is both a conscious and unconscious bastion of new identities. Looking at the metaverse from this perspective, the metaverse gravitates around the creation of a universe in which the digital parasitically inhabits users inhabiting this environment (Saker & Frith, 2022).

The process of agency is also something to consider. Agency refers to a user's ability to exercise his or her will. Thus, it can be argued that, in the metaverse, the process of agency is one by which the user from the real-world projects his or her

Table 4.1 Aspects of selves (Lloyd, 2019)

Actual self	The attributes one believes they actually possess. A realistic appraisal of one self.
Ideal self	The attributes one would like to ideally possess; i.e. one's hopes, aspirations, and wishes.
Ought self	The attributes one believes they should possess; i.e. a representation of one's duties, obligations, and responsibilities.
Personal self	The attributes one believes they possess which separate them from others and gives them a distinct personality.
Social self	The attributes of how one sees them self. How one evaluates and thinks about one self.
Ideal social self	The attributes of how one would like others to see in them. The image one creates for one self.
Looking-glass self	The attributes one believes they possess through how they see themselves reflected in the reactions from others.
Possible self	The attributes one may become in the future; i.e. outcomes which are hoped for or even dreaded.
Digital, Online or Cyber self	The attributes one presents to others, in a social sense in an online environment.

self onto the virtualization. By way of incarnational agency, people in the physical world make decisions, take actions, and express their self-will in the digital world, thereby producing a personal narrative that is the source of legitimacy for the metaverse to be contributed to by anyone. It not only establishes the virtual self as a legitimate interactant (e.g., through the display of a million faces, through an avatar, etc.), but it also forges interplays and overlaps in an unparalleled distributed network (Hu & Liu, 2022).

Self-presentation: Impression Management

For Erving Goffman (1959), a Canadian-born US sociologist, impression management is a method of gaining approval during interactions, which is often a condition for prosperous social life (as an alternative to a selfishly motived one). Impression management is an example of self-presentation. Self-presentation is a key factor of one's self-concept; the self is influenced by this process of self-presentation. The image displayed by a person to others also depends on his or her social status, social class, and/or resources. Impression management is analogized to a theatrical play in which daily self-presentation is compared to stage acting. For Goffman, the person performing on the stage conveys a message to an audience in the frontstage area and then retreats to the backstage area where he or she acts more like his or her true self. During this frontstage performance, the player engages in self-presentation and it is the backstage moment that is considered a moment "where the performer can reliably expect that no member of the audience will intrude" (Goffman, 1959, p. 114). This comparison between backstage and frontstage performance has been effortlessly transferred to the metaverse, with a number of studies suggesting that the virtual self is merely a frontstage persona (especially during interpersonal encounters).

The Expansive Approach vs. the Restrictive Approach

Impression management can be contemplated from two approaches: the expansive approach and the restrictive approach. From the expansive perspective, impression management is viewed as reflecting normal social behavior (Hogan, 1983); it is perceived as a process to allow others to derive conclusions based on how the information is presented (Schlenker, 1986). Here, self-presentation is to be considered as a strong trait of social behavior, one which is a keystone to the structure of social life (Schlenker, 1980). Impression management consists of packaging information in a manner intended to lead others to certain conclusions (Schlenker & Weigold, 1989).

When using the restrictive perspective, impression management and self-presentation are inherently linked to motives that are seen to be self-centered, such as those in instances where someone is "on a power trip," which leads to suggestions of deceitful or self-aggrandizing behavior. When analyzing impression management through the lens of a restrictive approach, self-presentation is believed to be enacted for the purpose of gaining an immediate audience (real) or an audience that will emerge in the immediate future (Schlenker & Weigold, 1992). Here, self-presentation is to be viewed as a specific behavior that is only enacted under certain conditions, or it is employed only by a certain group of people (Jones & Pittman, 1982).

Implications for the Metaverse

The reality that people in the metaverse tend to present themselves differently from how they do it in the physical world can be a challenge. In the metaverse, self-presentation is usually more performative because of (1) the high level of physical and emotional engagement made possible by virtual platforms and (2) the technological affordances not available offline (Heilweil, 2021). Although impression management is only one aspect of interpersonal interactions in the metaverse, it is a central one because it underscores the user's ability to conveniently change his or her appearance, behavior, and overall identity. Bente et al. (2008) already addressed the users' calculated emphasis on online identity features as one of the massive motivators of frequent computer-mediated communication. Based on the sender–receiver model, both interactants in the metaverse construct an imaginary audience as a result of the ways in which individuals adapt their self-presentation from the image they want to convey to their audience; it is also highly dependent on context and social cues (Litt & Hargittai, 2016).

Self-discrepancy Theory

Attestations to self-representation through avatars can be explained by both impression management theory (Mummendey, 1995) and self-discrepancy theory (Strauman, 1996). Based on the tenets of impression management theory, people seek to present themselves as ideally and positively as possible (Reyers & Matusitz, 2012). In the same vein, self-discrepancy theory distinguishes the actual self from the ideal self. Whereas the former represents the current self-image, the latter consists of a representation of how a person would ideally like to be and the traits that one aspires to acquire. Negative reactions become apparent when a significant discrepancy between the ideal self and the actual self emerges. This explains why we endeavor to minimize this discrepancy through our use of avatars (Bessière et al., 2007). The hypothesis for this idealized virtual self also affirms that people display idealized aspects of themselves in digital networks in ways that the world should see them. Both impression management theory and self-discrepancy theory promulgate a general idealization of avatars to achieve as much positivity in self-presentation as possible (Manago et al., 2008).

Avatar: Descriptions

In the metaverse today, the pinnacle of self-presentation and impression management is the avatar. The term comes from a Sanskrit word that symbolizes the incarnation of a deity (Park et al., 2021). Looking at it even more deeply, "avatar" comes from *avatāra*, a Sanskrit word that signifies "descent." In Hinduism, it is an embodiment of a deity in bodily form on Earth—e.g., like a divine teacher. More precisely, it means "an incarnation, embodiment, or manifestation of a person or idea" (cited in Kuru, 2023, p. 2). In contemporary times, an avatar is any effort by a user to create an individualized persona, generally one with visible traits. This grants them the license to manufacture a virtual representation of themselves that may or may not reflect their physical appearance offline (Sykownik et al., 2022).

Implications for Communication

Avatars are more and more pervasive because of their representations in virtual environments. They allow their users to interact and communicate at will. An increasingly large component of computer-based communication is done (or assisted) by avatars (Zimmermann et al., 2023). Scholarship (e.g., De Larios & Lang, 2014; Lee & Leeson, 2015) indicates avatars' astonishing ability to fulfill the need for expressing and displaying one's self online. In addition, avatars influence not only the kinds of communication channels available for self-disclosures but also the type of self-disclosure itself (in direct or indirect ways). In fact, the possibility to create an avatar that highly resembles a person in real life (with respect to appearance and psychological/behavioral attributes) may encourage or hamper self-disclosure based on those very appearance and psychological/behavioral attributes (Hooi & Cho, 2014).

Avatar-mediated communication can yield several benefits over current forms of communication. First, in contrast to text-based communication, avatar-mediated communication provides more possibilities for nonverbal interactions and body language—such as gestures, movements, and facial expressions (akin to those in face-to-face encounters) (Takano & Tsunoda, 2019). Users can also participate in proxemic interactions with others in order to convey supportive emotions and intimacy (Green-Hamann et al., 2011). Lastly, with full immersion in an interactive environment devised for diverse types of social activities, users can experience a uniquely profound social presence and emotional interdependence with others (van der Land et al., 2011), which constitute a crucial apparatus for effective communication in a mediated environment (Biocca et al., 2003).

These possibilities can add up-to-the-minute communicative effects to virtual 3D worlds. One specific illustration is the ability of users to transfer their own emotions or brain activities to their avatars. Another illustration is the impressive prospect for users to convert their body language into quasi-identical avatar communication. Through such flawless-looking body gestures, emotions, and moods, users can now tell a story through oral communication or even through art performance and dancing. To this last point, avatars can enact the sound–motion mappings of human dancers, assisted by deep learning, and replicate dancing movements with precise degrees of dance-likeness and emotional expressivity (Wallace et al., 2021).

Implications for the Long Term

Overall, there is sufficient evidence to believe that avatars not only improve online communication, but they can also change how people interact with others. Behavioral confirmation helps describe the pathway for this change. Behavioral confirmation is the process by which the expectations of one individual (usually, the "perceiver") cause another individual (usually, the "target") to behave in a manner that meets the perceiver's expectations (Snyder et al., 1977). A person's choice of an avatar leads to better alternate means of self-expression and communication; they will contribute to that person's "real" self-concept as he or she acts on his or her preferences for appearance and self-expression. The myriad possibilities to experiment with avatars represent a privilege that enables people to assume characteristics not usually present or available to any specific individual in real life (Montemorano, 2020).

The future of a metaverse focusing on an increasingly sophisticated avatar-based environment corroborates the power of avatars and their representation. In the next stages, the metaverse could become an autonomous, self-sustaining virtual world that operates in concert with the three-dimensional world. Under these circumstances, the avatars will not only be the embodiment of human users in the physical world, but they will also perform various activities in real-time owing to an unlimited number of users—possibly simultaneously in multiple virtual worlds. If, indeed, the metaverse can provide interoperability between platforms and through heterogeneous virtual worlds, users will create and widely share content across these worlds, particularly for games (e.g., Minecraft) and transfer such content to other platforms or games (e.g., Roblox) with a continuously renewing identity and experience. Even more, virtual platforms can be interconnected with our material world through a multitude of channels. How can it be done? In addition to avatars, users can access or distribute information through head-mounted helmets or displays, by using mobile headsets (e.g., Microsoft Hololens), by relying on computer agents that interact with smart devices and robots—just to name a few (Lee et al., 2021).

Case Study 1: Ready Player Me

"Ready Player Me" is a cross-game avatar platform available to the creators and residents of the metaverse. More specifically, it is a web-based cross-platform avatar game that can build a personalized avatar from a single selfie. This avatar can be included in any game and other applications involving human characters. The user can download the avatars in a standardized file format free of charge. A popular one is GLB, which stands for GL Transmission Format Binary file (Dhanjan & Steed, 2021). With Ready Player Me, signed-up members can control multiple avatars in the network and connect them with partner applications. The aforementioned builders, creators, and residents of the metaverse are tasked with eliminating the virtual walls to erect a more open and connected metaverse (Altundas & Karaarslan, 2023). For example, Ready Player Me avatars can be incorporated in the system of blendshapes that allow the detection and display of facial expressions or facial features. Each value of a blendshape is determined in proportion to the value of the various key points of the face (Le Tarnec et al., 2022). The possibilities for high-definition images, integrated into anything that the user wishes, reflect the magnitude of our identities online. It symbolizes the negation of the bodily self from the text (Davidson, 2022).

The sentiment of individual freedom can evidently be felt in Ready Player Me, which invites one to "level up your virtual experience with selfie-based avatars that people love and emotionally connect with" (see https://readyplayer.me/). The glorification of the selfie is reprocessed into the created avatar. It does so by codifying identity within the two dimensions, negating the third dimension, while the avatar implicitly involves the addition of that third—thereby negating the dimensionality of images and validating Baudrillard's theory of simulacra as a destructive energy. In essence, the selfie is a picture, still and emotional like a painting. However, the avatar makes it three-dimensional from the processing of lines of text (code) by the digital tools. It could be argued that the selfie merely consists of codes when digitized, but printing it out gives the aura of an authentic, analogue copy. Conversely, printing the avatar only becomes a snapshot—a selfie of the avatar—or simply lines of a code (as a representation of the architecture). The selfie of the avatar is not the reproduction of the person any more. Instead, it embodies the hypervirtual and alludes to a future situation of analogue pictures of identities that are exclusively virtual (Davidson, 2022).

Case Study 2: Second Life: Avatars in Place

Somewhat analogous to Ready Player Me, "Second Life" is also an online multimedia platform that enables users to manufacture an avatar, use it to interact with others, and share user-created content within a multiplayer metaverse (Irwin et al., 2023). As a 3D virtual environment, it gives users a place where they can be represented by anything they desire, be it an online persona or object. Second Life has enjoyed massive attention. Users can use it to attend social events such as concerts, press conferences, and classes, as well as to buy land, clothes, and gadgets, or even to visit their friends and relatives. In a specific case, in February 2022, SM Culture Universe (SMCU), an Asian entertainment company, developed a metaverse collaboration with a firm called "The Sandbox," which was one of the most supreme metaverse platforms in the world. The objective of this partnership was to host events in the metaverse, such as concerts and fan gatherings, and to produce various games and other digital products (Wong et al., 2022).

Liboriussen (2012) looked at Second Life as a way in which people create their avatars in order to inhabit a place. For Merleau-Ponty (1964), the concept of place is highly dependent on our body, or rather, on a feeling of bodily grounded agency. As Casey (1997) points out, Merleau-Ponty also bestows the concept of place "a virtual dimension overlooked in previous accounts. A place I inhabit by my body [...] I come to a place as providing an indefinite horizon of my possible actions." For a user in Second Life, this "virtual dimension" of "possible actions" means that even when a building is discovered or examined from a camera view (i.e., with an avatar-disembodied, high-rising camera), the avatar remains important in the scene. Its presence is experienced as a virtual avatar, implying that the user's feeling of proportion is enlightened by the out-of-sight avatar. Whether the building is tall or low, huge or small, it is always measured against the avatar. A slope becomes walk-constraining or walk-affording based on how the avatar is visualized to move up it. The key nuance, here, makes sense when discussing the concept of inhabitation (in the Casey quote above: "A place I inhabit") (Liboriussen, 2012, pp. 33–46).

What inhabitation emphasizes is that identifying with a place through a sense of bodily presence and agency transcends the ability to enact a set of well-defined actions. This can be interpreted as a definition of the avatar's role in digital world experiences (remember the notion of "my virtual body") (Liboriussen, 2012, pp. 33–46). Even in other Second Life settings, the body can remain a constitutive entity. In a nutshell, with Second Life, in order to make successful avatar appearances, users need to rely on their virtual bodies/identities because they are capable of inhabiting even the most seemingly remote or vacuous places of the virtual platform. Inasmuch as a location/site/setting/scene is a "possible habitat" for any "body," it can count as a place (Casey, 1997, p. 235).

In a similar "possible habitat" situation, Merleau-Ponty calls upon his "virtual body" to "inhabit" any location/site/setting/scene that users feel like inhabiting. Architect and architectural theorist Juhani Pallasmaa (2005) puts forward an idea similar to Merleau-Ponty's: "In an environmental experience, there is an unconscious bodily identification with the object, a projection of the body pattern onto what is experienced, or a physical mimesis, an unconscious mimicry" (p. 75). Such projection of body into an online multimedia platform such as Second Life is important not only for the experience of place in general but also in specific professions such as architecture. As such, in the twenty-first century, when architects use avatars in their models, sketches, and other representations of unrealized projects, the projections acquire a graphical focus and a more direct bodily foundation. The avatar replaces the otherwise purely imagined virtual body. It becomes a tool that assists in the inhabitation of virtual place by proxy (Liboriussen, 2012).

In the aforementioned example of "possible habitat," Second Life abounds with images of homes, where it is amazingly painless to locate groves, huts, fireplaces, and canopied beds. All of these features appear flat (if we would just think of the computer screen), but less flat when the embodied and avatar-based character of user interaction is considered. For instance, a canopied bed is felt not only on a screen but also through and with an avatar, thereby upholding some of its attributes as, say, a miniature home within the home with a certain degree of privacy (Liboriussen, 2012). In the metaverse, a bed is occupied by proxy when the avatar is used on the participant's behalf. Therefore, it is not totally true to make broad statements such as "[i]n the virtual world the existence of a 'building' is purely symbolic" (Harry et al., 2008, p. 65). What we perceive as symbolically pure is contaminated by the user's sense of agency. This scenario grants some degree of "dwelling-centered life" on digital images, even if it is precisely an interactive simulation of a dwelling—not the real, offline thing (Liboriussen, 2012, pp. 33–46).

References

Adoni, H., & Mane, S. (1984). Media and the social construction of reality. *Communication Research, 11*(3), 323–340. https://doi.org/10.1177/009365084011003001

Ahn, S. J., Kim, J., & Kim, J. (2022). The bifold triadic relationships framework: A theoretical primer for advertising research in the metaverse. *Journal of Advertising, 51*(5), 1–16. https://doi.org/10.1080/00913367.2022.2111729

Ahn, S. J., Kim, J., & Kim, J. (2023). The future of advertising research in virtual, augmented, and extended realities. *International Journal of Advertising, 42*(1), 162–170. https://doi.org/10.1080/02650487.2022.2137316

Altundas, S., & Karaarslan, E. (2023). Cross-platform and personalized avatars in the metaverse: Ready player me case. In E. Karaarslan, Ö. Aydin, Ü. Cali, & M. Challenger (Eds.), *Digital twin driven intelligent systems and emerging metaverse* (pp. 317–330). Springer.

Barnlund, D. (1970). A transactional model of communication. In J. Akin, A. Goldberg, G. Myers, & J. Stewart (Eds.), *Language behavior: A book of readings in communication* (pp. 43–61). De Gruyter Mouton.

Belk, R. W. (1988). Possessions and the extended self. *Journal of Consumer Research, 15*(2), 139–168. https://doi.org/10.1086/209154

Belk, R. W. (2013). Extended self in a digital world. *Journal of Consumer Research, 40*(3), 477–500. https://doi.org/10.1086/671052

Belk, R. W. (2016). Extended self and the digital world. *Current Opinion in Psychology, 10*, 50–54. https://doi.org/10.1016/j.copsyc.2015.11.003

Benford, R. D., & Hunt, S. A. (1992). Dramaturgy and social movements: The social construction and communication of power. *Sociological Inquiry, 62*(1), 36–55. https://doi.org/10.1111/j.1475-682X.1992.tb00182.x

Bente, G., Krämer, N. C., & Eschenburg, F. (2008). Is there anybody out there? Analyzing the effects of embodiment and nonverbal behavior in avatar-mediated communication. In E. A. Konijn, S. Utz, M. Tanis, & S. B. Barnes (Eds.), *Mediated interpersonal communication* (pp. 131–157). Routledge.

Berger, P. L., & Luckmann, T. (1966). *The social construction of reality: A treatise in the sociology of knowledge*. Anchor Books.

Berger, P. L., & Luckmann, T. (1991). *The social construction of reality (New Edition)*. Penguin Books.

Bessière, K., Fleming Seay, A., & Kiesler, S. (2007). The ideal elf: Identity exploration in world of warcraft. *Cyberpsychology & Behavior, 10*(4), 530–535. https://doi.org/10.1089/cpb.2007.9994

Biocca, F., Harms, C., & Burgoon, J. K. (2003). Toward a more robust theory and measure of social presence: Review and suggested criteria. *Presence: Teleoperators & Virtual Environments, 12*(5), 456–480. https://doi.org/10.1162/105474603322761270

Boothby, E. J., Clark, M. S., & Bargh, J. A. (2014). Shared experiences are amplified. *Psychological Science, 5*(12), 2209–2216. https://doi.org/10.1177/0956797614551162

Brewer, M. B., & Gardner, W. (1986). Who is this 'we'? Levels of collective identity and self representations. *Journal of Personality and Social Psychology, 71*(1), 83–93.

Burns, R. A., Crisp, D. A., & Burns, R. B. (2018). Competence and affect dimensions of self-concept among higher education students: A factorial validation study of an academic subject-specific self-concept. *European Journal of Psychology of Education, 33*(4), 649–663.

Burr, V., & Dick, P. (2017). Social constructionism. In B. Gough (Ed.), *The Palgrave handbook of critical social psychology* (pp. 59–80). Springer.

Carey, J. W. (1989). *Communication as culture: Essays on media and society*. Routledge.

Casey, E. (1997). *The fate of place: A philosophical history*. University of California Press.

Christie, B. (1985). *Human factors of information technology in the office*. Wiley.

Collett, J. L. (2011). What kind of mother am I? Impression management and the social construction of motherhood. *Symbolic Interaction, 29*(3), 327–347. https://doi.org/10.1525/si.2005.28.3.327

Cooley, C. (1902/1998). *On self and social organization*. University of Chicago Press.
Craig, R. T. (2007). Pragmatism in the field of communication theory. *Communication Theory, 17*(2), 125–145. https://doi.org/10.1111/j.1468-2885.2007.00292.x
Dhanjan, J. S., & Steed, A. (2021). Revisiting the Scene-Graph-as-Bus concept: Inter-networking heterogeneous applications using glTF fragments. In *IEEE conference on virtual reality and 3D user interfaces abstracts and workshops* (p. 20676968). https://doi.org/10.1109/VRW52623.2021.00068
Davidson, K. J. (2022). *Semiotic modelling of identity and communication in virtual reality, augmented reality, and mixed reality*. University of Tartu.
De Larios, M., & Lang, J. T. (2014). Pluralistic ignorance in virtually assembled peers: The case of world of warcraft. *Games and Culture, 9*(2), 102–121. https://doi.org/10.1177/1555412013512894
Dennis, A. R., & Kinney, S. T. (1998). Testing media richness theory in the new media: The effects of cues, feedback, and task equivocality. *Information Systems Research, 9*(3), 256–274. https://doi.org/10.1287/isre.9.3.256
Eisenberg, E. M. (1984). Ambiguity as strategy in organizational communication. *Communication Monographs, 51*(3), 227–242. https://doi.org/10.1080/03637758409390197
Eisenberg, E. M. (2008). The social construction of healthcare teams. In C. P. Nemeth (Ed.), *Improving healthcare team communication: Building on lessons from aviation and aerospace* (pp. 9–22). Ashgate.
Eisenberg, E. M., Goodall, H. L., Jr., & Tretheway, A. (2007). *Organizational communication: Balancing creativity and constraint* (5th ed.). St. Martin's Press.
Fulk, J., Schmitz, J., & Steinfield, C. W. (1990). A social influence model of technology use. In J. Fulk & C. W. Steinfield (Eds.), *Organizations and communication technology* (pp. 117–142). Sage.
García-Jiménez, L., Rodrigo-Alsina, M., & Pineda, A. (2017). The social construction of intercultural communication: A Delphi study. *Journal of Intercultural Studies, 38*(2), 22844. https://doi.org/10.1080/07256868.2017.1289903
Garfinkel, H. (1967). *Studies in ethnomethodology*. Prentice-Hall.
Goffman, E. (1959). *The presentation of self in everyday life*. Doubleday.
Green-Hamann, S., Eichhorn, K. C., & Sherblom, J. C. (2011). An exploration of why people participate in second life social support groups. *Journal of Computer-Mediated Communication, 16*(4), 465–491. https://doi.org/10.1111/j.1083-6101.2011.01543.x
Grieve, G., & Heston, K. (2012). Finding liquid salvation: Using the Cardean ethnographic method to document second life residents and religious cloud communities. In N. Zagalo, L. Morgado, & A. Boa-Ventura (Eds.), *Virtual worlds and metaverse platforms: New communication and identity paradigms* (pp. 288–306). IGI Global.
Haraway, D. (2006). A cyborg manifesto: Science, technology, and socialist-feminism in the late twentieth century. In J. Weiss, J. Nolan, J. Hunsinger, & P. Trifonas (Eds.), *International handbook of virtual learning environments* (pp. 117–158). Springer.
Harry, D., Offenhuber, D., & Donath, J. (2008). Function follows form: The social role of virtual architecture. In S. Doesinger (Ed.), *Space between people: How the virtual changes physical architecture* (pp. 64–70). Prestel.
Heilweil, R. (2021, November 3). Facebook is backing away from facial recognition. Meta Isn't. *Vox.* Retrieved on February 23, 2023 from https://www.vox.com/recode/22761598/facebook-facial-recognition-meta
Higgins, E. T. (1987). Self-discrepancy: A theory relating self and affect. *Psychological Review, 94*(3), 319–340. https://doi.org/10.1037/0033-295X.94.3.319
Hogan, R. (1983). A Socioanalytic theory of personality. In M. M. Page (Ed.), *Nebraska symposium on motivation* (pp. 55–89). University of Nebraska Press.
Hogg, M. A., Terry, D. J., & White, K. M. (1995). A tale of two theories: A critical comparison of identity theory with social identity theory. *Social Psychology Quarterly, 58*(4), 255–269. https://doi.org/10.2307/2787127

Hooi, R., & Cho, H. (2014). Avatar-driven self-disclosure: The virtual me is the actual me. *Computers in Human Behavior, 39*, 20–28. https://doi.org/10.1016/j.chb.2014.06.019
Hu, Y., & Liu, C. (2022). The 'Metaverse society': Beyond the discourse intrinsic potential and transformative impact. *Asia Pacific Academy of Science, 3*(2), 1–14. https://doi.org/10.54517/met.v3i2.2128
Irwin, P., Crepinsek, M., & Coutts, R. (2023). The use of avatars: Challenging longstanding approaches for experiential learning in nursing. *Interactive Learning Environments*. https://doi.org/10.1080/10494820.2022.2109176
Jones, E. E., & Pittman, T. S. (1982). Toward a general theory of strategic self-presentation. *Psychological Perspectives on the Self, 1*(1), 231–262.
Kuru, K. (2023). MetaOmniCity: Towards urban metaverse cyberspaces using immersive smart city digital twins. *IEEE Access, 11*, 1–26. https://doi.org/10.1109/ACCESS.2023.3272890
Lee, B. W., & Leeson, P. R. C. (2015). Online gaming in the context of social anxiety. *Psychology of Addictive Behaviors, 29*(2), 473–482. https://doi.org/10.1037/adb0000070
Lee, L.-H., Braud, T., Zhou, P., Wang, L., Dianlei, X., Lin, Z., Kumar, A., Bermejo, C., & Hui, P. (2021). All one needs to know about metaverse: A complete survey on technological singularity, virtual ecosystem, and research agenda. *Journal of Latex Class Files, 14*(8), 1–65.
Le Tarnec, H., Augereau, O., Bevacqua, E., & De Loor, P. (2022). Improving collaborative learning in virtual reality with facial expressions. In *ACM international joint conference on pervasive and ubiquitous computing and the 2022 ACM international symposium on wearable computers* (pp. 303–307). https://doi.org/10.1145/3544793.3560387
Liboriussen, B. (2012). Collective building projects in second life: User motives and strategies explained from an architectural and ethnographic perspective. In N. Zagalo, L. Morgado, & A. Boa-Ventura (Eds.), *Virtual worlds and metaverse platforms: New communication and identity paradigms* (pp. 33–46). IGI Global.
Litt, E., & Hargittai, E. (2016). The Imagined Audience on Social Network Sites. *Social Media + Society, 2*(1), 1–12. https://doi.org/10.1177/2056305116633482
Lloyd, N. M. (2019). Identity construction in online dating. University of Otago, Dunedin, New Zealand: Unpublished Doctoral Dissertation.
Manago, A. M., Graham, M. B., Greenfield, P. M., & Salimkhan, G. (2008). Self-presentation and gender on MySpace. *Journal of Applied Developmental Psychology, 29*(6), 446–458. https://doi.org/10.1016/j.appdev.2008.07.001
Markus, H., & Nurius, P. (1986). Possible Selves. *American Psychologist, 41*(9), 954–969. https://doi.org/10.1037/0003-066X.41.9.954
Martin, T. E., Jr. (2017). Framing leadership: The social construction of leadership within the academic field of communication studies. *Cogent Social Sciences, 3*(1), 1328794. https://doi.org/10.1080/23311886.2017.1328794
Marzaleh, M. A., Peyravi, M., & Shaygani, F. (2022). A revolution in health: Opportunities and challenges of the metaverse. *EXCLI Journal, 21*, 791–792. https://doi.org/10.17179/excli2022-5017
Mead, G. H. (1934). *Mind, self, and society from the standpoint of a social behaviorist*. University of Chicago Press.
Merleau-Ponty, M. (1964). *The visible and the invisible* (Trans. Alphonso Lingis). Northwestern University Press.
Miranda, S. M., & Saunders, C. S. (2003). The social construction of meaning: An alternative perspective on information sharing. *Information Systems Research, 14*(1), 87–106. https://doi.org/10.1287/isre.14.1.87.14765
Montemorano, C. (2020). Body language: Avatars, identity formation, and communicative interaction in VRChat. *Student Research Submissions, 361*. Retrieved on June 9, 2023 from https://scholar.umw.edu/student_research/361/
Moretti, G., & Schlemmer, E. (2012). Virtual learning communities of practice in metaverse. In N. Zagalo, L. Morgado, & A. Boa-Ventura (Eds.), *Virtual worlds and metaverse platforms: New communication and identity paradigms* (pp. 149–165). IGI Global.

Mumby, D. K. (1989). Ideology & the social construction of meaning: A communication perspective. *Communication Quarterly, 37*(4), 291–304. https://doi.org/10.1080/01463378909385551

Mummendey, H. D. (1995). *Psychologie der Selbstdarstellung [Psychology of self-representation]*. Hogrefe.

Pallasmaa, J. (2005). *The eyes of the skin: Architecture and the senses*. Wiley.

Park, S., Kim, S. P., & Whang, M. (2021). Individual's social perception of virtual avatars embodied with their habitual facial expressions and facial appearance. *Sensors, 21*(17), 5986. https://doi.org/10.3390/s21175986

Parmentier, G., & Rolland, S. (2009). Consumers in virtual worlds: Identity building and consuming experience in second life. *Recherche et Applications en Marketing, 24*(3), 43–55. https://doi.org/10.1177/205157070902400302

Pearce, W. B. (1995). A sailing guide for social constructionists. In W. Leeds-Hurwitz (Ed.), *Social approaches to communication* (pp. 88–113). Guilford Press.

Quadir, A., & Vijayakumar, V. (2019). Dynamic ranking of cloud services for web-based cloud communities: Efficient algorithm for rating-based discovery and multi-level ranking of cloud services. *International Journal of Web Based Communities, 15*(3), 248–270. https://doi.org/10.1504/IJWBC.2019.101811

Reyers, A., & Matusitz, J. (2012). Emotional regulation at Walt Disney World: An impression management view. *Journal of Workplace Behavioral Health, 27*(3), 139–159. https://doi.org/10.1080/15555240.2012.701167

Robinson, L. (2007). The cyberself: The self-ing project goes online, symbolic interaction in the digital age. *New Media & Society, 9*(1), 93–110. https://doi.org/10.1177/1461444807072216

Rogers, C. R. (1959). A theory of therapy, personality and interpersonal relationships as developed in the client-centered framework. In S. Koch (Ed.), *Psychology: A study of a science* (pp. 184–256). McGraw Hill.

Rusbult, C. E., Finkel, E. J., & Kumashiro, M. (2009). The Michelangelo phenomenon. *Current Directions in Psychological Science, 18*(6), 305–309. https://doi.org/10.1111/j.1467-8721.2009.0165

Saker, M., & Frith, J. (2019). From hybrid space to dislocated space: Mobile virtual reality and a third stage of mobile media theory. *New Media & Society, 21*(1), 214–228. https://doi.org/10.1177/1461444818792407

Saker, M., & Frith, J. (2022). Contiguous identities: The virtual self in the supposed metaverse. *First Monday, 27*(3). https://doi.org/10.5210/fm.v27i3.12471

Saura, J. R., & Bennett, D. (2019). A three-stage methodological process of data text mining: A UGC business intelligence analysis. *Symmetry-Basel, 11*(4), 519. https://doi.org/10.3390/sym11040519

Schlenker, B. R. (1980). *Impression management: The self-concept, social identity, and interpersonal relations*. Monterey, CA.

Schlenker, B. R. (1986). Self-identification: Toward an integration of the private and public self. In R. Baumeister (Ed.), *Public self and private self* (pp. 21–62). Springer.

Schlenker, B. R., & Weigold, M. F. (1989). Goals and the self-identification process: Constructing desired identities. In L. A. Pervin (Ed.), *Goal concepts in personality and social psychology* (pp. 243–290). Lawrence Erlbaum.

Schlenker, B. R., & Weigold, M. F. (1992). Interpersonal processes involving impression regulation and management. *Annual Review of Psychology, 43*(1), 133–168. https://doi.org/10.1146/annurev.ps.43.020192.001025

Schütz, A. (1967). *Phenomenology of the social world*. Northwestern University Press.

Short, J., Williams, E., & Christie, B. (1976). *The social psychology of telecommunications*. Wiley.

Shteynberg, G. (2015). Shared attention. *Perspectives on Psychological Science, 10*(5), 579–590. https://doi.org/10.1177/1745691615589104

Snow, D. A., Burke Rochford, E., Jr., Worden, S. K., & Benford, R. D. (1986). Frame alignment process, micromobilization, and movement participation. *American Sociological Review, 51*(4), 464–481. https://doi.org/10.2307/2095581

Snyder, M., Tanke, E. D., & Berscheid, E. (1977). Social perception and interpersonal behavior: On the self-fulfilling nature of social stereotypes. *Journal of Personality and Social Psychology, 35*(9), 656–666. https://doi.org/10.1037/0022-3514.35.9.656

Strauman, T. J. (1996). Stability within the self: A longitudinal study of the structural implications of self-discrepancy theory. *Journal of Personality and Social Psychology, 71*(6), 1142–1153. https://doi.org/10.1037/0022-3514.71.6.1142

Sykownik, P., Maloney, D., Guo, F., & Masuch, M. (2022). Something personal from the metaverse: Goals, topics, and contextual factors of self-disclosure in commercial social VR. *CHI '22: Proceedings of the 2022 CHI Conference on Human Factors in Computing Systems, 632*, 1–17. https://doi.org/10.1145/3491102.3502008

Takano, M., & Tsunoda, T. (2019). Self-disclosure of bullying experiences and social support in avatar communication: Analysis of verbal and nonverbal communications. *Proceedings of the International AAAI Conference on Web and Social Media, 13*, 473–481.

van der Land, S., Schouten, A., van den Hooff, & Feldberg, F. (2011). Modeling the metaverse: A theoretical model of effective team collaboration in 3D virtual environments. *Journal of Virtual Worlds Research, 4*(3).

Wallace, B., Martin, C. P., Tørresen, J., & Nymoen, K. (2021). Learning embodied sound-motion mappings: Evaluating AI-generated dance improvisation. *Proceedings of the 13th ACM SIGCHI Conference on C&C, 43*, 1–9. https://doi.org/10.1145/3450741.3465245

Waugh, R. F. (2000). Self-concept: Multidimensional or multi-faceted, unidimensional? *Education Research and Perspectives, 27*(2), 75–94.

Wentworth, W. M. (1980). *Context and understanding*. Elsevier.

Williams, D. E., & Page, M. M. (1989). A multi-dimensional measure of Maslow's hierarchy of needs. *Journal of Research in Personality, 23*(2), 192–213. https://doi.org/10.1016/0092-6566(89)90023-8

Wong, E. S., Wahab, N. H. A., Saeed, F., & Alharbi, N. (2022). 360-degree video bandwidth reduction: technique and approaches comprehensive review. *Applied Sciences, 12*, 7581. https://doi.org/10.3390/app12157581

Zhao, S. (2005). The digital self: Through the looking glass of telecopresent others. *Symbolic Interaction, 28*(3), 387–405. https://doi.org/10.1525/si.2005.28.3.387

Zimmermann, D., Wehler, A., & Kaspar, K. (2023). Self-representation through avatars in digital environments. *Current Psychology, 42*(3), 21775–21789. https://doi.org/10.1007/s12144-022-03232-6

Chapter 5
Nonverbal Communication in the Metaverse

Human communication consists of both nonverbal and verbal information. Nonverbal information is thought to convey as much as 93% of information in interpersonal encounters. One theoretical downside to traditional computer-mediated communication is the lack of this nonverbal information that is naturally exchanged by interactants. The missing nonverbal cues must be satisfactorily compensated for in order to achieve life-like communications—something akin to face-to-face communication (Matusitz, 2017). Facial expressions are the personification of nonverbal communication channels (Ekman, 2007). The face is the most recognizable sign and has unique features that "speak" on behalf of a person (Sproull et al., 1996). We are born with an innate power to sense and perceive the most influential individual (i.e., the mother) from the cradle. Infants can already discriminate facial features at just 2 months after birth, and they also favor facial expressions over other shapes and forms (Morton & Johnson, 1991). Concealing one's face goes hand in hand with concealing one's identity. For instance, wearing a facial mask when speaking may be considered a negative point in interpersonal communication (Diener et al., 1976).

The Importance of Nonverbals in the Metaverse

In the metaverse, the social and nonverbal communication cues with which we are familiar in the material world can be mimicked by modern technologies and platforms; they can confer a high level of perceived social presence, as would be in avatar-based interactions. This decreases the perceived uncertainty occasioned by the absence of social cues. As such, to make up for the missing nonverbal signals in avatars, many platforms provide manual control functions for the behavior and appearance of constructed personas. During dyadic social interactions, users can now absorb the others' intentions and emotions (Patterson, 1995). This is crucial because nonverbal cues are a keystone to communication and trust (Kasper-Fuehrer & Ashkanasy, 2001).

J. Matusitz, J. Dacas, *Communication in the Metaverse*, Signals and Communication Technology, https://doi.org/10.1007/978-3-031-63485-7_5

Prior Research

Prior research (e.g., see Bacharach & Gambetta, 1997) suggests that the totality of nonverbal cues—i.e., emotions, facial expressions, physical appearance, posture, gestures, and body language—are key to the development of trust. Even in avatar-based interactions, nonverbal communication signals such as clapping, laughing, and dancing are central to successful interpersonal encounters. Avatars are designed in such a way as to change one's gaze or bodily position to indicate the direction in which users are looking. This can be easily manipulated to engage others or to direct attention to a subject of interest. Studies have shown that avatars have the ability to perform gestures that replicate normal human nonverbal communications in team dynamics (Owens & Khazanchi, 2018). On the other hand, unlike the metaverse, digital systems that only afford a "camera on and camera off" option limit the choices that users have for visual representation. Overall, this restricts their faculty or aptness to communicate nonverbal information through body language—such as head movements for agreement or disagreement. Without proper access to nonverbal communication channels, users are less able to interact to a fuller extent or they have the sentiment of being excluded from a team/social group. In fact, doing away with just a few essential nonverbal cues completely alters the emotional content so necessary to communicate in the metaverse (Panda et al., 2022).

In their study, Chung et al. (2022) arranged nonverbal behaviors into three categories, inspired by research conducted by Maloney et al. (2020). The three categories are social behavior, individual behavior, and disturbing behavior. Social behavior and disturbing behavior could be defined as behaviors driven by favorable or disagreeable social interactions. In contrast, individual behaviors would be more difficult to define as behaviors with social meaning in themselves, although they might contribute to social behaviors in the long term. The results in Chung, Jo, and Lee's study reveal that, in the initial stage, active behavior is higher in the metaverse; individual behavior—more specifically, nonverbal behavior interacting with an object—is higher in a reality-based setting because of the affective responses according to that setting. Metaverse platforms show freedom to users, while reality-based spaces indicate more familiarity. Thus, the contextual meaning in each setting influences specific behaviors in a social virtual reality (VR) platform (Chung et al., 2022).

The Voice

The voice as a mode of expressing personality is not only natural, but it is also highly effective. The qualities of a participant's voice can be used as well-grounded indicators of personality in the metaverse (Jensen, 2016). It can give insights into some personality traits, such as extraversion, even with minimal short-term information (Levesque & Kenny, 1993), while other traits, such as openness and conscientiousness, require longer or repeated exposure to a person's communication (Goldberg, 1990). Prior research gives important cues on character personality expression through the paralinguistics of the voice. Examples include the type of dialogue, intonation, and so on. In the metaverse, modern technologies can combine human-like and machine-like characteristics, increasing expectations of life-like personalities and allowing them to align with an ideal range of personality expressions (Thomas et al., 2022).

Even before the rapid development of the metaverse, voice chat was already considered an efficient way of communicating, such as teleconference calls. Whoever was speaking at a specific point in time could be indicated visually in the setting, allowing the user to "make the connection"; that is, to associate an avatar with a voice. On the other hand, with teleconference calls, one voice often dominated, especially in large gatherings. Today, state-of-the-art VR platforms offer additional methods of communication to everyone in their avatar's vicinity (Satar & Özdener, 2008). For example, virtual worlds provide voice capabilities that assist in the exchange of 3D objects or artifacts. They also integrate the elements of unhampered face-to-face interaction (e.g., physical presence, the affordance to see and hear others, and concurrent communication) (Owens & Khazanchi, 2018).

Embodied Social Presence

Embodied social presence rests on the premise that (1) our body is at the center of communication and (2) an embodied persona, such as an avatar, influences the perceptions of users by getting them involved in a higher level of cognitive engagement during their shared activities and interactions (Mennecke et al., 2011). With embodied social presence, we can shift our focus between the digital and real self. In the metaverse, the vast majority of nonverbal communication cues are filtered through this embodied representation. When a person uses a virtual persona, the body representing him- or herself is expected to be seen as a life-like person. Hence, that representation will have an impact on the audience's perceptions of him or her and his or her actions associated with that representation (Biocca, 1997).

Digital Humans

Digital humans are such new category of embodied social presence. In essence, they are embodied conversational agents gifted with hyperrealistic, human-like features that rely on artificial intelligence (AI) to create interactions with others. They often use live neural networks to forge classifications of users' emotional states during encounters and respond by using a mix of adequate speech, facial behaviors, and body language (Ruppel et al., 2016). Digital humans are progressively being used in platforms for health communication, as well as education, work collaborations, and customer service. In order to make digital humans effective in these roles, it is imperative to get them to autonomously engage with others in an empathetic manner (Loveys et al., 2022).

In the metaverse, a great many situations and settings include people, which combines the social with the digital. When people socialize in a chat room or social media sites, text and images are visible to others, especially when those others are posting information or pictures (Ponder & Matusitz, 2017). When users socialize in the metaverse, it is embodied social presence because it is their avatars that are more visible. The spatial signals of communication operate in a way analogous to the real world. When X turns his head toward Y, Z can see the sidelong glance in real time. This nonverbal behavior is fundamental to the metaverse experience and explains why platforms such as VRChat are so prominent. In the context of VR, these nonverbal cues are highly persuasive. As such, Guadagno et al. (2007) have proved that simple modifications to avatars in a scene can lead to significant differences in persuasive outcomes. More precisely, avatars with high behavioral realism will have more eye contact. This means that they will more likely induce agreement to a proposed policy than avatars with lower behavioral realism. Likewise, avatars with close interpersonal distance between users induce more desired reactions than those who are farther away (Blascovich, 2022).

For Optimal Communication

As one can see, by virtue of 3D avatars, digital humans have the advantage of using a wide spectrum of nonverbal communication cues and interactive behaviors that add value to embodied social presence. Furthermore, as communication gives an aura of high behavioral realism, it seems that it occurs in real time, through instant messaging, with high-powered voice communication. A positive corollary is that it reduces invisibility between interactants and increase solipsistic introjection (Behm-Morawitz & Schipper, 2016). Solipsistic introjection is a phenomenon whereby a user perceives a voice or an image of another (in his or her mind) during virtual encounters (Mueller-Coyne et al., 2022). At the same time, mediated communication through the metaverse can contribute to a feeling of distance from others. Explained differently, a feeling of anonymity and imagination can emerge and cultivate toxic disinhibition (Behm-Morawitz & Schipper, 2016). Nevertheless, by showing appropriate interactive communication, users can become more efficient communicators (Skalski & Tamborini, 2007).

Some of the long-established components of nonverbal communication, those that are the most important, are facial expressions, hand gestures, arm movements, and posture because they allow us to express ideas, feelings, or messages. When conceiving realistic virtual avatars in domains such as learning, applying gestures will help infuse an emotional impression in the learners' embodied social presence (Yang & Ryu, 2021). AI can address most of these issues, for example by forecasting where our eyes should glance (even when blinking), to make the best rendering of life-like communication. This is optimal because it provides the most immerse experience. As times goes by, it will be vital for next-generation digital communications such as holographic light fields, which require this (Sivasankar, 2022).

In line with these contentions, the concept of digital humans can be seen through the use of digital pens. Today, they can help us write and draw a three-dimensional fashion, which would not be possible in real life. These pens are important in VRChat because they facilitate free communication in large groups of nonverbal users. This is a circle of individuals often referred to as "mutes." Mutes are users in VRChat that meet as a group for various reasons. They do not communicate orally (therefore, staying muted). Instead, they use other means such as gesticulations, simplified versions of sign language, those digital pens. A common feature in virtual worlds, these pens give rise to new forms of communication in the metaverse. It can happen through text produced anywhere, hanging mid-air, or on a wall (Asshoff, 2022).

Body Language in the Metaverse

To transform communitive intent into congruous body language, it would be useful to first understand what proper body language signifies. By and large, body language is a form of nonverbal communication that a person (or animal) uses through mimicry, gestures, and movements. Ideal body language is one that is meaningful when acknowledged in the subconscious of the audience. In fact, it is an ability to communicate feelings and thoughts to the receiver in a clear and more precise manner (Yurtsever & Eken, 2022). This matter has been systematically explored not only in the discipline of communication but also in psychology, sociology, anthropology, business, and the health and sports sciences. These academic branches have demonstrated that body language is precious when considered from various perspectives (Müller et al., 2013).

Reasons for Body Language

The following are some of the motives to justify why body language is of utmost importance in both offline and metaverse communication. To begin, body language allows interactants to be better understood or appreciated. It is particularly crucial during first encounters because the old age of "making first impressions" not only rings true, but it can also be best shown through body language. On many occasions, it communicates what words cannot. There are different forms of body language, including, but not limited to, (1) naturally facial or mimicking movements, (2) eye and brow movements, (3) physique, hand, and foot postures, (4) proxemics (the study of interpersonal distance), and (5) gestures at large (Rowlands, 2011). Through any of these five forms, exaggeration of body language can be helpful when other means of expressions to get points across are not available. Of course, other ways to show emotions or intent can be used. Examples include moving fast on one's chair with one's arms in the air, in a way mimicking how to show excitement (i.e., not being able to stand still or indicating a lot of positive energy). Other common uses of body language could be pointing with one's index and waving at people to get their attention (Asshoff, 2022).

Replacing Face-to-Face or Verbal Exchanges?

Our everyday interactions are evolving and are commensurate with the expansion of the metaverse. A new digital body language could even emerge and replace our familiar analog, face-to-face exchanges (Dhawan, 2021). Computers progress not only at recognizing gestures but also at creating new ones. They enable us to connect with other users more naturally and, in due course, comprehend and understand an unparalleled range of emotions and body movements. To cite one example, considering that the photoreceptors in our eyes are densest in the fovea—i.e., a region in the retina where visual acuity is highest—a metaverse technology could soon enhance our natural abilities to the highest resolution possible, thereby enhancing our perception in the process (perchance everything else in our peripheral vision too). Eye-tracking is another visual capability for supreme immersive interface: VR could offer the most reliable information through eye-tracking by getting our eyes to be highly focused on one zone (Sivasankar, 2022).

Nevertheless, in spite of the aforementioned existing qualities and possible enhancements in the future, body language in the metaverse will likely not fully replace face-to-face or verbal language. To be certain, the metaverse equipment already opens venues for exploration of body language never witnessed before, and it will facilitate our determination to understand or evaluate people's behaviors (Beck et al., 2013). Certain gestures will also allow us to take advantage of others' social intuition and communicative skills (Roth et al., 2015). Nevertheless, in avatar-based communication, body language requires a natural language primary model because spoken language will still remain essential for normal interactions. By extension, a great deal of body language in VR will remain unconscious by the user, even when virtually presented, and it will still be constructed through an artificial code complemented by verbal language. Because body language in the metaverse is, in part, instinctual, it cannot be flawless. Spatializing play, for example, is a secondary model system via an avatar that still requires verbal language—or at least the fundamentals of it. Thus, we are compelled to ask, again, whether it is conceivable to have such communication without verbal language (Davidson, 2022).

Incongruences Between Real and Virtual

Most users in the metaverse expect an online persona to behave in a manner congruent with its appearance and any major discrepancy in behavior would be considered disturbing. There are multiple reasons why mismatching can surface; it could also be a problem related to body movements or the motion of the avatar in general. To this point, the styles in which we move can be influenced by different factors, both physical and psychological. In the metaverse, if the physical or emotional state of the online persona is not befitting the captured actor, this may engender negative reactions (conscious or unconscious) on the part of the viewer, believing that there is something off-center or out of character (McDonnell, 2012). Naturally, the incongruences between the real and virtual persona may be uncontrollable, perhaps caused by a mechanical glitch, or due to unconscious effects (of which neither the sender nor the receiver is aware). This phenomenon is actually explained by the Uncanny Valley (UV) theory. The UV theory is a standard for describing how virtual humans can appear unintentionally otherworldly or even scary, especially when they defy the users' expectations. The UV hypothesis was first laid out by robotics professor Masahiro Mori (1970).

Masahiro Mori

Initially, Mori prognosticated that, when robots look more human than expected, it will appear so human that observers will begin to find subtle imperfections problematic. Lately, the UV hypothesis has been applied to human avatars. For example, according to results related by McDonnell (2012), it is obvious that certain female mannerisms in digital games or graphics should not be applied to male characters (and vice versa) because audiences will judge this conflicting information as unnatural or eerie. Studies in experimental psychology have demonstrated how highly developed the human brain can be at detecting inconsistencies by detecting erroneous biological motions (and discerning between even the most subtle characteristics of these motions). Johansson (1973) noted that natural motion, when spatial information is lacking, can offer a substantial perception of a moving human character. His "point-light" displays were conceived in a way that distinguished biological motion information from other sources of information—i.e., those that are typically mixed with the motion of a human (such as form or outline).

However, McDonnell (2012) reported that form and motion can generate unintended consequences. To be specific, if the gender of the captured actor is incongruent with that of the virtual character, players will find this unsettling. Body shape had no influence on ratings of emotional body motions, meaning that any virtual body can be operated to convey emotion decisively. This was true for both cartoon characters and realistic body replica of the captured actors. Similarly, virtual faces integrated with motion capture are equally capable of transmitting subtle cues associated with lying (as humans do in the real world). Additionally, the rendering style did not affect the impression of the performance on the viewers. This implies that, in the case of strong motion cues, visual cues do not change participants' opinions.

Decoding Language

In a communicative environment in which interactants increasingly use human-like artificial agents, identifying inconsistencies is considered of utmost importance by some. AI can also assist in measuring individuals' reactions to content generated artificially (Guzman, 2019). Donath (2021) opines that AI systems are improving in a meteoric manner, and that they will get better and better at dealing with human complexity and emotions. For instance, devices such as electroencephalography are already excellent (though not completely) at evaluating our responses to media stimuli (due to nonverbal cues such as noise) (Pickering, 2010). Recent advances in functional magnetic resonance imaging (MRI) scanners (Turner et al., 2019) or in brain recording indicate that, in the near future, it will be easier to decode language through noninvasive means (Tang et al., 2022). Big-tech companies (e.g., Meta) and those in the digital communication industry are already designing noninvasive technologies that can track brain activity and decode language, based on said electroencephalograms (and other forms of magnetoencephalography). They can do so by applying models using standard deep convolutional networks (Défossez et al., 2022).

Channels of Feedback

During nonverbal communication encounters in the metaverse, users can provide four different channels of feedback. These are (1) visual channels and auditory channels, (2) haptic channels, (3) olfactory channels, and (4) gustatory channels. To begin, visual channels and auditory channels refer to feedback created by the sense of sight and sound. Visual feedback is generated by changes in user action, usually appearing as a stereoscopic viewpoint through smart glasses or headsets. Auditory feedback can be classified into two different types: sound effect, allowing users to immerse themselves better in the metaverse, and voice, an important type of acoustic feedback that enables users to communicate with an avatar because of more senses of reality (Zhao et al., 2022).

Haptic channels refer to feedback created by the sense of touch. Here, users can feel various materials, textures, and temperatures, or even shapes and patterns with their fingertips. They are also made capable of perceiving thickness, dampness, smoothness, pressure, vibration, or friction. Haptic channels involve different force types: tension, traction, reaction, resistance, and impact, which assist in improving the user's experience in the metaverse. Gloves or exoskeletons are useful at simulating shapes (with the user's hands) or energizing other haptic qualities such as stiffness, friction, or slippage (Bouzbib et al., 2021). Olfactory channels refer to feedback created by the sense of smell. The main conduit of olfactory feedback is the dispersion of different scents to match the requirements of specific interaction tasks. Olfactory displays can also synchronize odorants from a digital designation. The feedback from the olfactory channel can improve the users' immersive experience in the metaverse. An olfactory display is frequently made up of a palette of odorants, a flow delivery apparatus, and a control algorithm that establishes the blending ratios, concentration, and timing of the stimulus (Richard et al., 2006).

Finally, gustatory channels refer to feedback created by the sense of taste. This can be triggered in two ways: (1) through a chemical combination that sprays into the user's mouth or to a small region where the user can lick with his or her tongue, or (2) through digital taste actuation technology that generates multiple taste sensations. Tools can be devised to achieve the process of gustatory stimulation (with or without chemicals). As such, the "Virtual cocoon" can spray flavors straight into the user's mouth. The "Food Simulator" employs chemical and mechanical associations to reproduce food chewing sensations by giving chemical additives, biting power, chewing sound, and vibration to the user (Ranasinghe et al., 2011). Providing thermal stimulation to the tongue and arousing the TRPM5 (transient receptor potential cation channel subfamily M member 5) taste channel can magnify the relish of sweet, bitter, and umami tastes (Karunanayaka et al., 2018).

Avatar Manipulation

The appearance of the avatars that users create or with which they interact can be changed based on user preferences. Sometimes, avatars can even be manipulated unbeknownst to the user—e.g., using data that the user may have shared unquestionably or accidentally. In more specific cases, avatar manipulation happens through clever use (or misuse) of the users' nonverbals. Let us go back to the case study on eye-tracking described earlier in this chapter. Eye-tracking data can reveal information on the users' pupil dilation and gaze fixation in response to visual stimuli (Wang, 2011). It can also offer valuable insights into their emotional responses. Such nonverbal signals can contribute to intelligent machine learning and AI algorithms that, in turn, churn out personalized, idealized interactants within a minute of the avatar-based encounter in the metaverse (Bailenson et al., 2008). The same process of avatar manipulation can be actuated for motion data that seeks to provide a window into the users' proxemic behavior and body language cues (Fast, 1970). And the same can be said about physiological data based on electroencephalograms and skin conductance machines that produce levels of emotional activation (Horvers et al., 2021).

Voice Manipulation

In addition to avatar manipulation based on such outward appearance, voice can be easily tweaked or manufactured to be enticing to users. Vocal cloning software can maneuver vocal tone, fluctuation, pitch, timbre, pronunciation, articulation, and volume. Users can also switch their voice from male to female (and vice versa).

Appealing voices that are smooth in density, thickness, range, or width can be effortlessly created through auditory morphing (Bruckert et al., 2010). This type of software can also impersonate a particular person's voice. Vocal cloning in online personas has already been ascertained as highly influential because it can increase social influence and attraction (Torre et al., 2021). It is replete with a wide spectrum of cues that can be manipulated for various reasons, including attractiveness and any method that seeks to control interaction behaviors (Zibrek et al., 2020).

Multiplexity

Multiplexity consists of creating multiple profiles for impression management purposes, in which case a user will appear differently based on the type of audience (e.g., friends on Facebook vs. colleagues on LinkedIn) (Kane et al., 2014). In the metaverse, multiplexity is less common because users have the same profile across platforms, with the same avatar (that has the user's appearance). This also happens because of the interoperable nature of platforms, which may restrict our ability to set up multiple profiles. As such, it has been shown that several large providers have agreed to maintain a shared database (through blockchain technology) of registered users. At the same time, reduced use of multiplexity may cause users to be more careful or honest because anonymity is more difficult. The consequences of reduced anonymity can have wider and relevant implications (Marabelli & Newell, 2022).

Nevertheless, users relish the idea of multiplexity because it is designed to be appealing to them. It can even be argued that avatar manipulation through the creation of multiple profiles would make users more physically and psychologically complex (Ledbetter & Meisner, 2021). The personality of a computer-based agent could be transformed into the most likeable personality type that the user wants to exhibit (Liew & Tan, 2016). A more developed iteration of AI or a human-driven avatar could consist of identifying and highlighting a user's medical and mental conditions. Ultimately, this would foster a feeling of closeness and trust with that user. With all intents and purposes, although many applications of multiplexity through avatar manipulation happen for benevolent reasons, there could be cases driven by maliciousness too (Buck & McDonnell, 2022).

Anthropomorphism

Anthropomorphism is a practice in which human-like features are ascribed to an object or an animal. Users have exhibited a preference for anthropomorphic assistants over other types of agents (Kontogiorgos et al., 2019). Part of the reason is that turning an embodied conversational agent into a more human-like one is a visual stimulus to interactants in the metaverse (Seeger et al., 2021). Additionally, there are other reasons. In terms of social cues, an anthropomorphic agent can assist in increasing the user's positive perceptions based on oral, auditory, visual, and even unconscious cues (e.g., response time) (Feine et al., 2019). Oral cues usually pertain to content-related cues (e.g., greetings) or style (e.g., formal or informal ways of speaking), whereas auditory cues have to do with pitch, tempo, volume, and so forth. Most of these social cues can be added with little effort, as would be the case when adding a response delay or conferring a person's name to the embodied conversational agent (Diederich et al., 2020).

These social cues can enhance the positive perception of anthropomorphism and elicit social responses. They are helpful when attempting to express friendliness and emotions. Anthropomorphism is increasingly used to establish emotional bonds because digital humans easily manipulate friendly styles and social–emotional cues to assure levels of trustworthiness and interactivity (Zierau et al., 2020). However, it should be remarked that nonverbal cues increase positive outcomes only when added in conjunction with verbal cues and humanlikeness (Seeger et al., 2021). Nonverbal cues need verbal ones for optimal user engagement and satisfaction (in order to heighten friendliness and perceived empathy). Nonverbal signals alone can certainly boost a service worker's perceived competence and friendliness (e.g., by adding a "thumbs-up" icon), but influencing trust and customer loyalty is more efficient when text is added somewhere (Gabbott & Hogg, 2000). Appeal can be secured when changing certain facial features, as it is cardinal to adhere to realistic placement and facial proportions when developing a human-like character. Eye size, for example, can induce trustworthiness: larger eyes are said to appear more trustworthy (as opposed to narrow eyes) (Zibrek et al., 2019).

Deep Synthesis

If a chief goal of avatar manipulation in the metaverse is to achieve identity creation (and, by the same token, recognition), then deep synthesis seeks to synthesize digital images based on the features of objects, including facial and bodily features, and mimic the features of the referent as authentically as possible (Li et al., 2022). This can be achieved by means of face reproduction, face synthesis, and speech synthesis. Face reproduction amounts to controlling the facial expression of the online persona—including the mouth, smiles, eyebrows, eyes, and head—to monopolize the facial expression of that persona. Dissimilar to face swap, face reproduction does not entail replacing the target character but altering the facial expression of a real individual and let him or her determine that, indeed, the changes have been successfully effectuated. Face synthesis consists of generating a new face virtual image akin to the real face (or even replacing some facial features for purposes of advertising and user account profiles). Speech synthesis is designed to produce a particular voice model to convert text information into tones that are similar to natural speech and rhythm (Liu & Zheng, 2022).

In this fashion, deep synthesis is a useful approach to construct a digital twin world, but it can be perceived as a "Trojan horse" when the tweaking of face recognition has malicious intent. Deep synthesis can set a paradigm shift in research and development in that the higher the optimization of face recognition, the higher the leap into applications to other domains. This shift to unchartered territories in the metaverse not only presents a challenge to the existing biometric recognition technologies, but it can also expand the development potential of AI application in the field of biological information to such an extent that many human professions could be useless in the future. Initially, deep synthesis was predominantly used in computer-generated imagery for film production, but in 2017 it crossed over into areas when it could be easily used to replace the heroines' face in US adult photos and films with renowned Hollywood actresses.

Deep synthesis is predicted to morph into full-body synthesis and digital virtual humans in unprecedented ways by the end of the decade. Among them is the already-existing practice of deepfake, which is commonly understood in the public realm as a form of face swapping. Put simply, face swapping consists of replacing the face of a person with that of another one. This was exemplified right above with the case of US adult photos and films. Typical software applications include ZAO and FakeApp, which can help generate the most basic face synthesis (though they cannot reach high-precision deep synthesis information) (Liu & Zheng, 2022). Not surprisingly, the use of deep synthesis has been deemed objectionable in certain domains and has been monitored and restricted by a great many governments around the world (Liu & Zheng, 2022). Having said that, deep synthesis represents only a fraction of digital image synthesizing. There are other methods for applying specific technology and computer graphics for creating or improving images.

Emotions in the Metaverse: A Deeper Look

Emotion denotes an elaborate mixture of experiences of consciousness, bodily sensations, and behavior, expressing personal substance of externally and/or internally driven changes. As the connective tissue in human interactions, emotions are also present in the digital world. When combined with gamification, more profound communication for the participant can be impactful for the group of players or the game itself (Dias et al., 2023). Comprehending how the expression of emotions shapes the perception of others is beneficial to the application of digital human technology in the metaverse. For instance, in a situation where an employee has to interact with a customer, positive emotions can reinforce the customer–employee bond through emotional contagion, i.e., a phenomenon whereby emotions are conveyed from one person to another deliberately or unknowingly (Hennig-Thurau et al., 2006).

Emotional Contagion

Emotional contagion transcends facial expressions. It often comes through in textual online comments and icons/memes (Bösch et al., 2018). Emotions can have a more powerful impact when perceived as authentic and sincere. Higher rapport, in turn, increases interpersonal satisfaction. In the domain of customer service, for example, companies have worked on developing digital humans that can ooze a wide range of emotions, raising the question of whether AI agents affect user emotions in similar ways (Bock et al., 2020). Although emotional contagion within digital humans has been ascertained to build rapport and trust, the impact of microexpressions—facial expressions that occur for a fleeting moment, usually a fraction of a second—needs further examination (Loveys et al., 2020).

Research also confirms that social presence magnifies the intensity of emotions that circulate in virtual environments, which could even trigger a higher desire for emotional contagion in the material world (Riva et al., 2007). Kim and Biocca (1997) hypothesized that presence is like transportation from both ends: arrival and departure. Arrival is the sensation of being present in the virtual world and departure is the sensation of separation from the physical world. This signals that the user, given the deeper sense of presence, momentarily flees the real world into the virtual world. This notion of social presence is the basic building block for effective metaverse applications. They could be easily utilized for persuasion because emotional contagion is a cogent factor in successful performance through human information processing and other cognitive variables.

Facial Expressions

Facial expressions belong to the kinesics classification of visual social cues and gestures. Kinesics, in and of itself, is the formal study of body language and movements (Matusitz, 2014). The face can forge three types of signals that lead to all sources of information (Adamo et al., 2019). These are static (long-term features, such as the contour of the face), rapid (muscle movements, such as facial expressions), and slow signals (facial transformations, such as wrinkles). Facial expressions are born out of a swift response system that allows humans to react promptly to events, whether impactful or not. The developed composition of our facial muscles signifies that humans benefit from a general collection of 40 facial actions to display these emotions—this does not include the thousands of possible microexpressions (Matsumoto & Willingham, 2009). Emotions can be further classified into two dimensions: (1) valence, going from positive to negative (or pleasant vs. unpleasant), and (2) arousal, going from active to passive (Horovitz & Mayer, 2021). Aware that humans have universal expressions, irrespective of culture, famous US psychologist Paul Ekman established a universal correlation between facial muscle movements and seven particular emotions (i.e., happiness, sadness, anger, fear, surprise, disgust, interest) (Ekman, 1970). In the face of this universality, individual differences emerge in the vigor of each muscle movement. Researchers also discovered some asymmetrical measures of facial regions that make individual differences firm and stable (Liu et al., 2003).

Emoticons

Emoticons are symbolic expressions comprising symbols such as ":", "=", "-", ")", or "(". They generally represent facial expressions. Emoticons can be read in multiple ways, such as ":-(" for a sad face or "()" for a happy one. Observing such emoticons in addition to texts is primordial in identifying genuine sentiments, such as happiness, exasperation, anger, grief, and so forth, because they can be soon interpreted as positive, negative, or neutral. With the rapid progress in technology, the fusion of real and virtual space evokes the need to investigate human sentiments about new interactions, events, or objects in the metaverse. In particular, the advent of AI has improved emotional online communication, where participants can share their own emotional issues on virtual platforms (Bhattacharyya et al., 2023).

Enhancing Communication in the Metaverse

As a rule of thumb, emoticons fall into a range of 14 emotions: worry, happiness, fear, joy, neutral, sadness, surprise, anger, relief, hate, disgust, enthusiasm, boredom, and shame (Midhan et al., 2023). In 2001, the term "emoticon" was included into most dictionaries and lexicons as a type of internet lingo (cited in Landa & Zhou, 2022). "Emote" is a diminutive word for "emotion" and evocative of the conventional ways of showing emotions in Western cultural contexts. When one "emotes" in the metaverse, one manipulates an avatar in a manner dictated or recommended by that avatar's internal code. Avatars made available by the platform's creators offer a static assortment of emoting capabilities named Wave, Clap, Point, Cheer, Dance, Backflip, Sadness, and Die (Montemorano, 2020).

Such emoting capabilities illustrate how emotions are dynamic and complex in nature, rarely experienced in an unadulterated form of a single emotion. Therefore, having at least 14 different emoticons is amply adequate for conveying the communication and expression of feelings accurately. They also allow users more personalization and adaptation in the metaverse. Considering that everyday emotional life and social communication can be strengthened by emoticons, their constant use and streamlining over time can provide a window into the participants' psychological well-being or status in a particular interaction. It can also give insights into their mental health or psychopathology. As emotion-related channels, emoticons can surface from numerous sources of affective information, especially (1) bodily reactions (e.g., facial expressions, gestures), (2) developments in physiological signals (e.g., oscillating electrical voltages in the brain, cortical hemodynamic activity, cardiovascular function, respiration rate, temperature, electrodermal activity), and (3) channels of communication and interaction (e.g., texts, images, video, and audio) (Dias et al., 2023).

The Benefits of Anonymity

The fact that communication in the metaverse can be anonymous enables participants to present their identities and express themselves in more unconventional ways, whether their approaches are real or fake. Anonymity certainly allows users to conceal the digital schizophrenia or paranoia from which they suffer in three-dimensional life. Both the physical and virtual worlds are more visual, meaning unequivocally that images are more important than words. Emoticons play a large role in that visual world, empowering users with new meanings and new modes of expression. Emoticons help construct a scene that diffuses meanings for viewers to read in ways that would be more complicated with words (Benavides-Vanegas, 2020). We now know that users can communicate through a multiplicity of channels, such as body language, gestures, and facial expressions. Thus, interactants in the metaverse can easily convey their sentiments with emoticons and other paralinguistic facets of language. From this vantage point, employing a 3D avatar that projects emotions in the metaverse is so helpful in providing an enhanced sense of social presence (Hwang et al., 2023).

Emojis

Emojis were developed in the late 1990s by NTT DoCoMo, a Japanese communications corporation. The name is a combination of the words *e* and *moji*, which means something like "pictograph." In comparison with emoticons, emojis are actually pictures, going from a set of painted nails (💅) to a slightly fanciful ghost (👻). Whereas emoticons are designed to represent emotions in text-prone environments, emojis are meant to add a veneer of emotional nuance to messages and, at the same time, be the character set used by many modern operating systems, Unicode. In fact, emojis are to be considered letters from a non-Western language, a bit like what Japanese and Chinese characters are. This also implies that the software has to categorically support them—otherwise it will display a placeholder icon, or simply a blank space (Hern, 2015).

As small pictograms, emojis are useful to represent larger concepts. They often become recurrent signs, from a larger discourse, that get reduced to a single image, song, or character model. Emojis can become memes, cultural units that are also highly limited in character space or symbolization and that replicate across both traditional media and virtual platforms (Woolford & Matusitz, 2011). Memes serve several functions, primarily acting as grammatical units within virtual linguistics to inject some level of emotion into topics or debates. Emojis and memes personify the character limit of twenty-first-century communication. Focusing only on emojis here, it can be argued that they rarely denote anything concretely because they are symbols. Although some may be paradigmatic to the specific action, the act itself is the one that becomes the symbol—e.g., the recent symbol of resistance represented by the "raised fist" emoji, as the one used by the Black Lives Matter campaign. There is nothing symbolic about a wave for the meaning of "goodbye," but the "wave" emoji itself can become emblematic of the human act. Of course, there should always be some level of understanding of cultural awareness of the action for the icon to be interpreted or used without aberration. In other words, the objective visual appearance contravenes the deliberate subjective, conceptual, interpretation (Davidson, 2022).

Enhancing Communication in the Metaverse

Emojis are a staple attribute of our digital interactions because they make up for the lack of nonverbal or paralinguistic cues in the metaverse. As explained by Gawne and McCulloch (2019), emojis in the metaverse are as important as tone of the voice and body movements in face-to-face communication. As further illustration that emojis are an integral part of human communication in the digital age, Apple has "over 50 different emoji to choose from," so "you can express yourself in any way you want" (see https://apps.apple.com/us/app/metamoji-metaverse-emoji/id1630248638). The mode of communication that emojis represent is affective in nature because they express speaker attitudes or tones. Conversely, emojis can also be used in a nonaffective manner. This usually happens when they are used as icons for verbal concepts, particularly when directly substituting for a word. By extension, emojis can be included for strictly decorative purposes; for example, when they furnish a fancy frame for messages or posts, so they can be highlighted (Song, 2022).

In most cases, users incorporate emojis to convey the emotions that underlie their text-based interactions to get their messages across, mainly for brief words. Any virtual platform that applies semantic-based context recognition structures may understand emojis and turn them into audio data. A significant number of image-based emojis depend on image-recognition algorithms in order to wholly understand the semantic context (Bhattacharyya et al., 2023). In addition to shoring up the community aspects, within a virtual platform, that erase cultural and geographical barriers, it can also make geographical, real-world, features of identity more prominent. More precisely, it enables users to nurture comradeship through the objective reality of offline culture, developing the anonymous virtuality in some recognizable fashion—e.g., an emoji pictured horizontally rather than vertically (Danesi, 2017). No matter what, emojis are aids that uphold the cohesiveness of an online community. They improve visual consolidation and simplification of communication among members (Davidson, 2022).

Case Study: Kansei

Kansei is a Japanese word that cannot be easily translated in English. Broadly defined, the concept refers to humans' sensibility, sensitivity, emotions, creativity, imagination, encouragement, enthusiasm, and intuitive thinking. To add more exactitude, *Kansei* is understood to be the impression that individuals get when they come into contact with artifacts such as emojis and emoticons. This view is contingent on the environment or situation, and is a function of our senses of touch, sight, hearing, smell, taste, and even cognition. Afterward, a complex mind pattern matures and gets stored in our brain, encompassing all the impressions experienced; in doing so, it lays the foundation for human behavior. In essence, emoticons are sensations caused by a person's *Kansei*. This is why it should be viewed as a component of *Kansei* (Edama et al., 2021).

The metaverse can underpin the perception of others' *Kansei* to stimulate new awareness in day-to-day routines. People's *Kansei* can be observed through their gazing behavior and emoticons flanked by scenery objects. Emoticons can underlie the reason and intention that a user has, and make a specific scene (or simply a particular object of the scenery) a sensible place. In the study conducted by Edama et al. (2021), 90% of the participants claimed that emoticons helped them guess other users' intentions. Nonetheless, 40% of these participants also said that different emoticons should also be used, implying that emoticons, in and of themselves, are not sufficiently adequate to express all users' intentions. This corroborates Fujisawa et al.'s (2022) statement that emoticons tend to be more polysemous than emojis; that is, they can carry multiple meanings or their meanings can change when included within sentences. Hence, it would be challenging to validate emotions and intentions through emoticons only. Under these circumstances, it would be useful for metaverse users to explore other options in order to more accurately understand or express *Kansei*.

It would still be easy for users in the metaverse to develop new awareness of others' *Kansei* even in unfamiliar settings. Part of the reason is that it is human nature to express strong interest in other people's lives or intentions through emotions. Strong interest can also be seen in products displayed on virtual platforms. Notwithstanding emotions as being exceedingly subjective and individualistic, Desmet and Hekkert (2007) exposed the correlation between product elements and customer expression. Products beneficial to users conjure up pleasant emotions. *Kansei* engineering is a highly popular and powerful product development method in the metaverse because it manipulates the consumers' emotional needs (Hartono & Chuan, 2011). *Kansei* is generally successful in gauging human emotions vis-à-vis products and methods such as designs of eyewear, popup box, and e-commerce websites (Lokman, 2013).

References

Adamo, N., Dib, H. N., & Villani, N. J. (2019). Animated agents' facial emotions: Does the agent design make a difference? In L. T. De Paolis & P. Bourdot (Eds.), *Augmented reality, virtual reality, and computer graphics* (pp. 10–25). Springer.

Asshoff, R. (2022). *Welcome to VRChat: An ethnographic study on embodiment and immersion in virtuality reality*. Stockholm University.

Bacharach, M., & Gambetta, D. (1997). Trust in Signs. In K. S. Cook (Ed.), *Trust in Society* (pp. 148–184). Russell Sage Foundation.

Bailenson, J. N., Pontikakis, E. D., Mauss, I. B., Gross, J. J., Jabon, M. E., Hutcherson, C. A. C., Nass, C., & John, O. (2008). Real-time classification of evoked emotions using facial feature tracking and physiological responses. *International Journal of Human-Computer Studies, 66*(5), 303–317. https://doi.org/10.1016/j.ijhcs.2007.10.011

Beck, S., Kunert, A., Kulik, A., & Froehlich, B. (2013). Immersive group-to-group telepresence. *IEEE Transactions on Visualization and Computer Graphics, 19*(4), 616–625. https://doi.org/10.1109/TVCG.2013.33

Behm-Morawitz, E., & Schipper, S. (2016). Sexing the Avatar: Gender, sexualization, and cyber-harassment in a virtual world. *Journal of Media Psychology, 28*(4), 161–174. https://doi.org/10.1027/1864-1105/a000152

Benavides-Vanegas, F. S. (2020). Emoticons, memes and cyberbullying: Gender equality in Colombia. *Social Semiotics, 30*(3), 328–343. https://doi.org/10.1080/10350330.2020.1731169

Bhattacharyya, M., Roy, A., Midya, S., Mitra, A., Ghosh, A., & Roy, S. (2023). An emoticon-based sentiment aggregation on metaverse related Tweets. In A. E. Hassanien, A. Haqiq, A. T. Azar, K. C. Santosh, M. A. Jabbar, A. Słowik, & P. Subashini (Eds.), *International conference on artificial intelligence and computer vision* (pp. 358–367). Springer.

Biocca, F. (1997). The cyborg's dilemma: Progressive embodiment in virtual environments. *Journal of Computer-Mediated Communication, 3*(2), JCMC324. https://doi.org/10.1111/j.1083-6101.1997.tb00070.x

Blascovich, J. (2022). Social influence within immersive virtual environments. In R. Schroeder (Ed.), *The social life of avatars: Presence and interaction in shared virtual environments* (pp. 127–145). Springer.

Bock, D. E., Wolter, J. S., & Ferrell, O. C. (2020). Artificial intelligence: Disrupting what we know about services. *Journal of Services Marketing, 34*(3), 317–334. https://doi.org/10.1108/JSM-01-2019-0047

Bösch, K., Müller, O., & Schneider, J. (2018). Emotional contagion through online newspapers. *European Conference on Information Systems*, 171. https://aisel.aisnet.org/ecis2018_rp/171

Bouzbib, E., Bailly, G., Haliyo, S., & Frey, P. (2021). Can I Touch This?': Survey of virtual reality interactions via haptic solutions: Revue de Littérature des Interactions en Réalité Virtuelle par le biais de Solutions Haptiques. *IHM '21: 32e Conférence Francophone sur l'Interaction Homme-Machine, 9*, 1–16. https://doi.org/10.1145/3450522.3451323

Bruckert, L., Bestelmeyer, P., Latinus, M., Rouger, J., Charest, I., Rousselet, G. A., Kawahara, H., & Belin, P. (2010). Vocal attractiveness increases by averaging. *Current Biology, 20*(2), 116–120. https://doi.org/10.1016/j.cub.2009.11.034

Buck, L. E., & McDonnell, R. (2022). Security and privacy in the metaverse: The threat of the digital human. In *Proceedings of CHI Conference on Human Factors in Computing Systems* (pp. 1–4).

Chung, S. J., Jo, H.-J., & Lee, H. (2022). A comparison of behaviours and responses towards different social VR environments in initial social interaction. *Archives of Design Research, 35*(3), 53–67. https://doi.org/10.15187/adr.2022.08.35.3.53

Danesi, M. (2017). *The semiotics of Emoji: The rise of visual language in the age of the internet*. Bloomsbury Academic.

Davidson, K. J. (2022). *Semiotic modelling of identity and communication in virtual reality, augmented reality, and mixed reality*. University of Tartu.

Défossez, A., Caucheteux, C., Rapin, J., Kabeli, O., & King, J. R. (2022). Decoding speech from non-invasive brain recordings. *arXiv* 5. https://doi.org/10.48550/arXiv.2208.12266

Desmet, P., & Hekkert, P. (2007). Framework of product experience. *International Journal of Design, 1*(1), 13–23.

Dhawan, E. (2021). *Digital body language: How to build trust and connection, no matter the distance*. St. Martin's Press.

Dias, S. B., Jelinek, H. F., & Hadjileontiadis, L. J. (2023). Multisensed emotions as adaptation controllers in human-to-serious NeuroGames communication. *IEEE Communications Magazine, 61*(10), 38–44. https://doi.org/10.1109/MCOM.001.2200828

Diederich, S., Brendel, A. B., & Kolbe, L. M. (2020). Designing anthropomorphic enterprise conversational agents. *Business & Information Systems Engineering, 62*, 193–209. https://doi.org/10.1007/s12599-020-00639-y

Diener, E., Fraser, S. C., Beaman, A. L., & Kelem, R. T. (1976). Effects of deindividuation variables on stealing among halloween trick-or-treaters. *Journal of Personality and Social Psychology, 33*(2), 178–183. https://doi.org/10.1037/0022-3514.33.2.178

Donath, J. (2021). Commentary: The ethical use of powerful words and persuasive machines. *Journal of Marketing, 85*(1), 160–162. https://doi.org/10.1177/0022242920973975

Edama, I., Ayedoun, E., Takenouchi, H., & Tokumaru, M. (2021). Kansei perception support system to promote daily life awareness. In C. Stephanidis, M. Antona, & S. Ntoa (Eds.), *HCI International 2021* (pp. 241–248). Springer.

Ekman, P. (1970). Universal Facial Expressions of Emotions. *California Mental Health Research Digest, 8*(4), 151–158.

Ekman, P. (2007). *Emotions Revealed, Second Edition: Recognizing faces and feelings to improve communication and emotional life*. Holt Paperbacks.

Fast, J. (1970). *Body Language*. Simon & Schuster.

Feine, J., Gnewuch, U., Morana, S., & Maedche, A. (2019). A taxonomy of social cues for conversational agents. *International Journal of Human-Computer Studies, 132*, 138–161. https://doi.org/10.1016/j.ijhcs.2019.07.009

Fujisawa, A., Matsumoto, K., Yoshida, M., & Kita, K. (2022). Emotion estimation method based on emoticon image features and distributed representations of sentences. *Applied Sciences, 12*(3), 1256. https://doi.org/10.3390/app12031256

Gabbott, M., & Hogg, G. (2000). An empirical investigation of the impact of non-verbal communication on service evaluation. *European Journal of Marketing, 34*(3), 384–398. https://doi.org/10.1108/03090560010311911

Gawne, L., & McCulloch, G. (2019). Emoji as digital gestures. *Language Internet, 17*(2), 10–21.

Goldberg, L. R. (1990). An alternative "description of personality": The Big-Five factor structure. *Journal of Personality and Social Psychology, 59*(6), 1216–1229. https://doi.org/10.1037/0022-3514.59.6.1216

Guadagno, R. E., Blascovich, J., Bailenson, J. N., & Mccall, C. (2007). Virtual humans and persuasion: The effects of agency and behavioral realism. *Media Psychology, 10*(1), 1–22. https://doi.org/10.1080/15213260701300865

Guzman, A. L. (2019). Voices in and of the machine: Source orientation toward mobile virtual assistants. *Computers in Human Behavior, 90*, 343–350. https://doi.org/10.1016/j.chb.2018.08.009

Hartono, M., & Chuan, T. K. (2011). How the Kano model contributes to Kansei engineering in services. *Ergonomics, 54*(11), 987–1004. https://doi.org/10.1080/00140139.2011.616229

Hennig-Thurau, T., Groth, M., Paul, M., & Gremler, D. D. (2006). Are all smiles created equal? How emotional contagion and emotional labor affect service relationships. *Journal of Marketing, 70*(3), 58–73. https://doi.org/10.1509/jmkg.70.3.58

Hern, A. (2015, February 6). Don't Know the Difference between Emoji and Emoticons? Let Me Explain. *Daily Mail*. Retrieved on June 1, 2023 from https://www.theguardian.com/technology/2015/feb/06/difference-between-emoji-and-emoticons-explained

Horovitz, T., & Mayer, R. E. (2021). Learning with human and virtual instructors who display happy or bored emotions in video lectures. *Computers in Human Behavior, 119*, 106724. https://doi.org/10.1016/j.chb.2021.106724

Horvers, A., Tombeng, N., Bosse, T., Lazonder, A. W., & Molenaar, I. (2021). Detecting emotions through electrodermal activity in learning contexts: A systematic review. *Sensors, 21*(23), 7869. https://doi.org/10.3390/s21237869

Hwang, Y., Shin, D., & Lee, H. (2023). Students' perception on immersive learning through 2D and 3D metaverse platforms. *Educational Technology Research and Development, 71*(2), 1687–1708. https://doi.org/10.1007/s11423-023-10238-9

Jensen, M. (2016). Personality traits and nonverbal communication patterns. *International Journal of Social Science Studies, 4*(5), 57–70. https://doi.org/10.11114/ijsss.v4i5.1451

Johansson, G. (1973). Visual perception of biological motion and a model for its analysis. *Perception & Psychophysics, 14*(2), 201–211. https://doi.org/10.3758/BF03212378

Kane, G. C., Alavi, M., Labianca, G. (J.), & Borgatti, S. P. (2014). What's different about social media networks? A framework and research agenda. *MIS Quarterly, 38*(1), 275–304. https://doi.org/10.25300/MISQ/2014/38.1.13

Karunanayaka, K., Johari, N., Hariri, S., Camelia, H., Bielawski, K. S., & Cheok, A. D. (2018). New thermal taste actuation technology for future multisensory virtual reality and internet. *IEEE Transactions on Visualization and Computer Graphics, 24*(4), 1496–1505. https://doi.org/10.1109/TVCG.2018.2794073

Kasper-Fuehrer, E. C., & Ashkanasy, N. M. (2001). Communicating trustworthiness and building trust in interorganizational virtual organizations. *Journal of Management, 27*(235), 235–254. https://doi.org/10.1177/014920630102700302

Kim, T., & Biocca, F. (1997). Telepresence via Television: Two dimensions of telepresence may have different connections to memory and persuasion. *Journal of Computer-Mediated Communication, 3*(2), JCMC325. https://doi.org/10.1111/j.1083-6101.1997.tb00073.x

Kontogiorgos, D., Pereira, A., Andersson, O., Koivisto, M., Rabal, E. G., Vartiainen, V., & Gustafson, J. (2019). The effects of anthropomorphism and non-verbal social behaviour in virtual assistants. *Proceedings of the ACM International Conference on Intelligent Virtual Agents, 19*, 133–140. https://doi.org/10.1145/3308532.3329466

Landa, N., & Zhou, S. (2022). Using Emoticons to Reduce Transactional Distance: Navigating the Contextual Complexities of COVID-19 Imposed Virtual Learning Spaces. In N. Amin & R. Dhunpath (Eds.), *(Re)Learning to Teach in Contexts of Crises* (pp. 169–193). CSSALL Publishers.

Ledbetter, A. M., & Meisner, C. (2021). Extending the personal branding affordances typology to parasocial interaction with public figures on social media: Social presence and media multiplexity as mediators. *Computers in Human Behavior, 115*(2), 106610. https://doi.org/10.1016/j.chb.2020.106610

Levesque, M. J., & Kenny, D. A. (1993). Accuracy of behavioral predictions at zero acquaintance: A social relations Analysis. *Journal of Personality and Social Psychology, 65*(6), 1178–1187. https://doi.org/10.1037/0022-3514.65.6.1178

Li, M., Wan, Y., & Gao, J. (2022). What drives the ethical acceptance of deep synthesis applications? A fuzzy set qualitative comparative analysis. *Computers in Human Behavior, 133*, 107286. https://doi.org/10.1016/j.chb.2022.107286

Liew, T. W., & Tan, S.-M. (2016). Virtual agents with personality: Adaptation of learner-agent personality in a virtual learning environment. *International Conference on Digital Information Management (ICDIM), 11*, 157–162.

Liu, Y., Schmidt, K. L., Cohn, J. F., & Mitra, S. (2003). Facial asymmetry quantification for expression invariant human identification. *Computer Vision and Image Understanding, 91*(1), 138–159. https://doi.org/10.1016/S1077-3142(03)00078-X

Liu, Z., & Zheng, Y. (2022). *AI ethics and governance: Black mirror and order*. Springer.

Lokman, A. M. (2013). KE as affective design methodology. *International Conference on Computer, Control, Informatics and Its Applications*, 14331090. https://doi.org/10.1109/IC3INA.2013.6819139

Loveys, K., Sebaratnam, G., Sagar, M., & Broadbent, E. (2020). The effect of design features on relationship quality with embodied conversational agents: A systematic review. *International Journal of Social Robotics, 12*(6), 1293–1312. https://doi.org/10.1007/s12369-020-00680-7

Loveys, K., Hiko, C., Sagar, M., Zhang, X., & Broadbent, E. (2022). I Felt Her Company': A qualitative study on factors affecting closeness and emotional support seeking with an embodied conversational agent. *International Journal of Human-Computer Studies, 160*(C), 102771. https://doi.org/10.1016/j.ijhcs.2021.102771

Maloney, D., Guo, F., & Wohn, D. Y. (2020). Talking without a Voice': Understanding Non-Verbal Communication in Social Virtual Reality. *Proceedings of the ACM on Human-Computer Interaction, 4*(CSCW2), 1–25. https://doi.org/10.1145/3415246

Marabelli, M., & Newell, S. (2022). Everything you always wanted to know about the metaverse* (*But Were Afraid to Ask). *Academy of Management Annual Meeting, 82*, 10–21.

Matsumoto, D., & Willingham, B. (2009). Spontaneous facial expressions of emotion of congenitally and noncongenitally blind individuals. *Journal of Personality and Social Psychology, 96*(1), 1–10. https://doi.org/10.1037/a0014037

Matusitz, J. (2014). Intercultural perspectives on cyberspace: An updated examination. *Journal of Human Behavior in the Social Environment, 24*(7), 713–724. https://doi.org/10.1080/10911359.2013.849223

Matusitz, J. (2017). The communication of suicide terrorism. In *Handbook of suicidal behavior* (pp. 159–177). Springer.

McDonnell, R. (2012). Appealing virtual humans. In M. Kallmann & K. Bekris (Eds.), *Motion in Games* (pp. 102–111). Springer. https://doi.org/10.1007/978-3-642-34710-8_10

Mennecke, B. E., Triplett, J. L., Hassall, L. M., Conde, Z. J., & Heer, R. (2011). An examination of a theory of embodied social presence in virtual worlds. *Decision Sciences Journal, 42*(2), 413–450. https://doi.org/10.1111/j.1540-5915.2011.00317.x

Midhan, T. M., Selvaraj, P., Raju, M. H. K., Reddy, B. P., & Bhaskar, T. (2023). Classification of mental health and emotion of human from text using machine learning approaches. *International Conference on Information Systems and Computer Networks, 6*, 23115373. https://doi.org/10.1109/ISCON57294.2023.10111973

Montemorano, C. (2020). Body language: Avatars, identity formation, and communicative interaction in VRChat. *Student Research Submissions, 361*. Retrieved on June 9, 2023 from https://scholar.umw.edu/student_research/361/

Mori, M. (1970). The Uncanny Valley. *IEEE Robotics & Automation Magazine, 19*(2), 98–100. https://doi.org/10.1109/MRA.2012.2192811

Morton, J., & Johnson, M. H. (1991). CONSPEC and CONLERN: A two-process theory of infant face recognition. *Psychological Review, 98*(2), 164–181. https://doi.org/10.1037/0033-295X.98.2.164

Mueller-Coyne, J., Voss, C., & Turner, K. (2022). The impact of loneliness on the six dimensions of online disinhibition. *Computers in Human Behavior Reports, 5*, 100169. https://doi.org/10.1016/j.chbr.2022.100169

Müller, C., Cienki, A., Fricke, E., Ladewig, S., McNeill, D., & Tessendorf, S. (2013). *Body–language–communication: An international handbook on multimodality in human interaction*. De Gruyter Mouton.

Owens, D., & Khazanchi, D. (2018). Exploring the impact of technology capabilities on trust in virtual teams. *American Journal of Business, 33*(4), 157–178. https://doi.org/10.1108/AJB-04-2017-0008

Panda, P., Nicholas, M. J., Gonzalez-Franco, M., Inkpen, K., Ofek, E., Cutler, R., Hinckley, K., & Lanier, J. (2022). AllTogether: Effect of avatars in mixed-modality conferencing environments. *2022 Symposium on Human-Computer Interaction for Work, Association for Computing Machinery, 8*, 1–10. https://doi.org/10.1145/3533406.3539658

Patterson, M. L. (1995). Invited article: A parallel process model of nonverbal communication. *Journal of Nonverbal Behavior, 19*(1), 3–29. https://doi.org/10.1007/BF02173410

Pickering, A. (2010). *The cybernetic brain*. University of Chicago Press.

Ponder, S., & Matusitz, J. (2017). Examining ISIS online recruitment through relational development theory. *Connections: The Quarterly Journal, 16*(4), 35–50. https://doi.org/10.11610/Connections.16.4.02

Ranasinghe, N., Karunanayaka, K., Cheok, A., Fernando, O., Nii, H., & Gopalakrishnakone, P. (2011). Digital taste and smell communication. *International ICST Conference on Body Area Networks, 6*, 78–84. https://doi.org/10.4108/icst.bodynets.2011.247067

Richard, E., Tijou, A., Richard, P., & Ferrier, J.-L. (2006). Multi-modal virtual environments for education with haptic and olfactory feedback. *Virtual Reality, 10*(3), 207–225. https://doi.org/10.1007/s10055-006-0040-8

Riva, G., Mantovani, F., Capideville, C. S., Preziosa, A., Morganti, F., Villani, D., Gaggioli, A., Botella, C., & Alcañiz, M. (2007). Affective interactions using virtual reality: The link between presence and emotions. *Cyberpsychology & Behavior, 10*(1), 45–56. https://doi.org/10.1089/cpb.2006.9993

Roth, D., Latoschik, M. E., Vogeley, K., & Bente, G. (2015). Hybrid avatar-agent technology—A conceptual step towards mediated 'Social' virtual reality and its respective challenges. *i-Com, 14*(2), 107–114. https://doi.org/10.1515/icom-2015-0030

Rowlands, M. (2011). *Body language: Representation in action*. MIT Press.

Ruppel, E. K., Gross, C., Stoll, A., Peck, B. S., Allen, M., & Kim, S.-Y. (2016). Reflecting on connecting: Meta-analysis of differences between computer-mediated and face-to-face self-disclosure. *Journal of Computer-Mediated Communication, 22*(1), 18–34. https://doi.org/10.1111/jcc4.12179

Satar, H. M., & Özdener, N. (2008). The effects of synchronous CMC on speaking proficiency and anxiety: Text versus voice chat. *The Modern Language Journal, 92*(4), 595–613. https://doi.org/10.1111/j.1540-4781.2008.00789.x

Seeger, A.-M., Pfeiffer, J., & Heinzl, A. (2021). Texting with human-like conversational agents: Designing for anthropomorphism. *Journal of the Association for Information Systems: JAIS, 22*(4), 931–967. https://doi.org/10.17705/1jais.00685

Sivasankar, G. A. (2022). Study of Blockchain technology, AI and digital networking in metaverse. *IRE Journals, 5*(8), 110–115.

Skalski, P., & Tamborini, R. (2007). The role of social presence in interactive agent-based persuasion. *Media Psychology, 10*(3), 385–413. https://doi.org/10.1080/15213260701533102

Song, Z. (2022). Interpersonal communication research in metaverse—Taking Si-Fi films as example. *Advances in Social Science, Education and Humanities Research, 664*, 2377–2382.

Sproull, L., Subramani, M., Kiesler, S., Walker, J. H., & Waters, K. (1996). When the Interface Is a Face. *Human–Computer Interaction, 11*(2), 97–124. https://doi.org/10.1207/s15327051hci1102_1

Tang, J., LeBel, A., Jain, S., & Huth, A. G. (2022). Semantic reconstruction of continuous language from non-invasive brain recordings. *bioRxiv* 4. https://doi.org/10.1101/2022.09.29.509744

Thomas, S., Ferstl, Y., McDonnell, R., & Ennis, C. (2022). Investigating how speech and animation realism influence the perceived personality of virtual characters and agents. *IEEE Conference on Virtual Reality and 3D User Interfaces (VR)*, 21724594. https://doi.org/10.1109/VR51125.2022.00018

Torre, I., Carrigan, E., Domijan, K., McDonnell, R., & Harte, N. (2021). The effect of audio-visual smiles on social influence in a cooperative human–agent interaction task. *ACM Transactions on Computer-Human Interaction (TOCHI), 28*(6), 1–38. https://doi.org/10.1145/3469232

Turner, B. O., Huskey, R., & Weber, R. (2019). Charting a future for fMRI in communication science. *Communication Methods and Measures, 13*(1), 1–18. https://doi.org/10.1080/19312458.2018.1520823

Wang, J. T.-Y. (2011). Pupil dilation and eye tracking. In M. Schulte-Mecklenbeck, A. Kuehberger, & J. G. Johnson (Eds.), *A handbook of process tracing methods for decision research: A critical review and user's guide* (pp. 185–204). Psychology Press.

Woolford, T., & Matusitz, J. (2011). The memetic engineering of anonymous, the cyberterrorist group. *International Journal of Cyber Warfare and Terrorism, 1*(4), 1–9. https://doi.org/10.4018/ijcwt.2011100101

Yang, E., & Ryu, J. (2021). Do we perceive emotional gender gesture of virtual avatar as intended? *International Journal of Educational Media and Technology, 15*(1), 129–139.

Yurtsever, M. M. E., & Eken, S. (2022). BabyPose: Real-time decoding of baby's non-verbal communication using 2D video-based pose estimation. *IEEE Sensors Journal, 22*(14), 13776–13784. https://doi.org/10.1109/JSEN.2022.3183502

Zhao, Y., Jiang, J., Chen, Y., Liu, R., Yang, Y., Xue, X., & Chen, S. (2022). Metaverse: Perspectives from graphics, interactions and visualization. *Visual Informatics, 6*(1), 56–67. https://doi.org/10.1016/j.visinf.2022.03.002

Zibrek, K., Martin, S., & McDonnell, R. (2019). Is photorealism important for perception of expressive virtual humans in virtual reality? *ACM Transactions on Applied Perception, 16*(3), 14. https://doi.org/10.1145/3349609

Zibrek, K., Niay, B., Olivier, A.-H., Hoyet, L., Pettré, J., & Mcdonnell, R. (2020). The effect of gender and attractiveness of motion on proximity in virtual reality. *ACM Transactions on Applied Perception (TAP), 17*(4), 1–15. https://doi.org/10.1145/3419985

Zierau, N., Engel, C., Söllner, M., & Leimeister, J. M. (2020). Trust in smart personal assistants: A systematic literature review and development of a research agenda. *International Conference on Wirtschaftsinformatik (WI), 15*. https://doi.org/10.30844/wi_2020_a7-zierau

Chapter 6
Interaction, Collaboration, and Training in the Workplace

On top of becoming a fully functional community of users who can engage in various forms of activities, the metaverse can conceivably transform people's day-to-day routines at their workplaces by developing opportunities for interaction, collaboration, and training. As a fundamentally social place, the metaverse expands the horizons for a wide variety of human activities through synergies and joint efforts (e.g., by interacting with colleagues and hosting virtual business meetings.)—and all this would be in addition to meeting the individual needs of employees (CBS News, 2021). Considering that the previous era already witnessed the technologization of business life for several decades, identifying up-to-the-minute virtual communication channels is worthy of attention because it has a tremendous impact on organizational productivity and success. Hence, it would be wise to adapt to increasingly new communication models in order to better communication for labor productivity and across-the-board benefits (Jamadi et al., 2022).

Of equal relevance is the reality that constant evolution of the external environment stimulates scholarly efforts in addressing digital transformations in the workplace. For example, the early 2020 breakout of the COVID-19 pandemic forced myriad corporations and institutions to impose limits on physical exchanges or interpersonal interactions among employees. As a result, these corporations and institutions had to guarantee productivity and performance by establishing activities through virtual platforms. It should come as no surprise that the metaverse will be constantly tapped for promoting employees' activities in virtual spaces or via 3D technologies. However, as many prior studies on the metaverse have focused on improving employees' content and services, additional research is needed to understand the metaverse's effects at the organizational level and, more precisely, its impact for interaction, collaboration, and training (Hwang et al., 2022).

The essence of the metaverse in the workplace can be captured through four features: realism, ubiquity, interoperability, and scalability. Realism means that the virtual world gives workers ample opportunity to feel psychologically and emotionally immersed. Ubiquity alludes to workers' connectedness to virtual platforms

J. Matusitz, J. Dacas, *Communication in the Metaverse*, Signals and Communication Technology, https://doi.org/10.1007/978-3-031-63485-7_6

with routine devices such as tablets, smartphones, and PCs. Interoperability refers to the capacity of computer systems or software to share and use information for all employees. The scalability of the metaverse comes through in networks that are pervasive and influential that they can host large swaths of users simultaneously and without facing recurrent technical problems (Mostajeran et al., 2022).

Estimated Value

Estimated values of the metaverse vary. Goldman Sachs (2022) believes that its value could reach US$ 12 trillion in the next decade, which could also comprise up to 33% of the current digital economy. On the other hand, Precedence, a Canadian and Indian-based research firm, predicts that, by 2030, the overall metaverse could be worth about US$ 1.6 trillion based on an estimated year-to-year growth rate of 50.74% (Global News Wire, 2022). Research by McKinsey and Company (2022) examined the potential of the metaverse to become the top marketing platform and stressed that virtual product sales through direct-to-avatar exchanges represent a market exceeding US$ 50 billion. In any case, organizations are providing a great deal of effort to develop their longer term strategies to determine how their business models and brands can excel in the metaverse. The announcement that Italy's first-tier soccer division, the Lega Nazionale Professionisti Serie A, is now using the Nemesis metaverse to screen matched between A-rated teams such as AC Milan and Fiorentina attests to the potential for new levels of media diffusion and brand awareness (Reuters, 2022). In addition, knowledge workers, that is, "workers whose input is knowledge resources to yield knowledge-based intellectual output" (Kianto et al., 2019, p. 179), have, to a certain extent, saved companies a lot of money because they have replaced traveling expenses, social gatherings, and 9–5 office work with their working-from-home practices and the like (Baptista et al., 2020).

In addition, the facts described above demonstrate that, as the metaverse has grown exponentially, so too has its user base. Corporations and institutions are seeking digital transformations to magnify their businesses so that participants in virtual platforms or those in the external environment—i.e., users who are not metaverse inhabitants yet—connect with such corporations and institutions. Thus, a multitude of entities are developing strategies to facilitate business transactions through digital technologies, even when experiencing all sorts of constraints as a result of supply chain and human resource utilization. The following sections focus on how the metaverse encourages smooth and well-planned interactions among organizational employees, reinforces homogeneity among coworkers (e.g., through heightened social presence), and promotes workplace satisfaction. It has been proved useful for virtual meetings, conferences, and corporate events because it prompts higher participation from members in organizations (Lee & Hwang, 2022).

Interaction and Collaboration

Interactivity is a rationale for circulating information within groups (Ball et al., 2021). In virtual reality (VR), work-related information, as well as social connection and supportive messages, can forge a sense of belonging in ways that offline infrastructure cannot (Kim et al., 2020). For example, bolstering up social interactions and online collaboration among workers in groups has been shown to increase both knowledge exchange and information-rich awareness as a whole (Martinez-Martinez et al., 2020). In a similar vein, when VR promotes interactivity between users, more collaborative innovation can be accomplished. Although employees are not necessarily more likely to reach a consensus through interactions in the metaverse, they are more likely to commit to the organization (i.e., some form of loyalty or allegiance) (Lee et al., 2016).

Collaboration is inherently linked to workplace communication. Its quality can be ensured through openness, clarity, truthfulness, and efficiency. With the metaverse, workplace communication improves in instances where interactants provide more feedback, linguistic diversity, multiple channels, and any technology that supports communication (Zhang et al., 2022). Communication practitioners constantly need to keep abreast with ways to engage effectively with participants across and within this new reality (Robbins, 2022). The metaverse has the benefit of giving employees the impression that another one is working with them or next to them. The strong feelings that can spring out of such digital interactions and collaborations are also shaped by our online identities and ability to interact within the virtual from multiple angles. The use of avatars for social presence in the workplace can significantly enhance the *esprit de corps* of organizational employees because the aforementioned "strong feelings" bring about more happiness, pride, fellowship, and common loyalty within a particular group and, thereby, foster more creative collaboration (Yfantis & Ntalianis, 2022).

Online Meetings and Virtual Offices

Online meetings in the metaverse give employees a virtual space to interact and collaborate on tasks, ranging from text-based chat rooms to fully immersive digital environments, and employing voice chat, instant messaging, or other modes of communication (Madsen & Matusitz, 2022). These virtual meetings are resourceful and flexible; they can serve the purpose of team collaboration, networking, keeping company, or making appearances at virtual events and conferences. Similarly, the use of virtual offices is a promising hallmark of the metaverse because it has the potential to restructure or make far-reaching changes in collaboration and work methods even more. By exploiting all the benefits of the metaverse's virtual environment, employees can cooperate in a customizable, shared digital fashion that offers additional benefits, such as lower overhead cost and recruitment of a global talent pool (Tariq et al., 2023).

Microsoft has endeavored to develop projects that can combine the mixed physical and digital worlds of Microsoft Mesh, a platform that enables participants to connect with a holographic presence, share space, and join forces with others from all four corners of the planet. A demo video has been made available by the big-tech corporation through the following link (https://learn.microsoft.com/it-it/mesh/overview). In particular, Microsoft Teams technologies has been successful in getting people to utilize ultimate messaging apps for their organizations—usually, through workspaces for live collaboration, communication, and participation in virtual meetings. Microsoft Teams technologies are specifically designed to stimulate interactive and collaborative work experience for remote workers or users scattered across the world. In 2021, Microsoft moved toward blending the physical and digital work worlds in an unusual fashion; that is, it began applying an approach that allows participants to use avatars in its Teams collaborative software. In another (brief) case study, the Hilton Hotel Group has resorted to using such digital tech to train human resources and personnel for guest management. In doing so, it will be possible to attribute human qualities to, say, "front office" AI. The hotel industry can turn chatbots—now commonplace in websites—into authentic interactions with "front office info points" directed by AI agents or digital humans (Caragnano, 2023).

Real-Time Multisensory Social Interactions (RMSIs) in the Workplace

Real-time multisensory social interactions (RMSIs) were discussed in detail in Chap. 3, but they have drawn the attention of global executives. As a brief reminder, RMSIs are defined as interactions between two or more participants that take place concurrently and entail multiple senses (e.g., sight, hearing, touch). Global executives conceive of the metaverse as an environment in which RMSIs will give consumers and corporations more advantages than two-dimensional (2D) computer-mediated platforms such as Zoom, Skype, and Discord. RMSIs accessed through VR headsets help organizations accomplish desired outcomes with respect to interaction performances (e.g., better solutions for work teams), positive evaluations (e.g., services provided to customers), and positive feelings of interactants (e.g., commitment to the workplace, satisfactory customer service) (Hennig-Thurau et al., 2022).

Three Outcome Categories

The value and efficiency of RMSIs can be ascertained by three basic interaction outcomes: the interactants' performance, their evaluations, and their emotional responses. Performance outcomes include the volume of creativity of work-team solutions (Brucks & Levav, 2022). Evaluation outcomes include perceptions of satisfactory customer service and are essential for RMSIs because they occur at the customer–frontline employee interface. Emotion outcomes usually pertain to the positive emotions of interactants and are fundamental to RMSIs because they contribute to unique consumption experiences, such as watching a movie together after work (Hennig-Thurau et al., 2022). All three outcome categories can greatly ameliorate the financial performance of firms. Performance outcomes, such as creativity of work-team solutions, can boost firms by directing their market and financial performance in unchartered territories (Im & Workman, 2004). So do evaluative metrics by consumers (e.g., service quality) and employees (e.g., job satisfaction) (Hogreve et al., 2017). The same applies to the emotions of the latter two (Hennig-Thurau et al., 2006).

RMSIs during Conference Meetings

A metaverse-based conference program today—which we might call SocialVR conference—generally uses session formats that differ in both arrangement (e.g., keynotes, workshops, presentations/exhibitions) and content (e.g., which could be more catered to the organization, to the employees, or the general public/consumers). The rooms will change based on sessions. Conventional session formats, such as presentations and keynotes, are still popular today. The typical metaverse-based conference structure looks like a discussion-oriented conference format (rather than a submission-driven arrangement). Interactions among participants are primarily via voice chat, using spatial audio, so that the direction and distance of the speakers are represented by the volume and direction of the sound. In most cases, specific gestures are supported; depending on the type of VR headset, head and hand movements will be more prominent, or certain finger gestures will be. Emoticons can also be incorporated in addition to avatars. Because computer screens, and even more so with VR headsets, can be tiring or draining over a long period, SocialVR conference meetings are set up to last for a maximum of 2 hours each day and up to 6 hours per week. To ensure a comprehensive and optimal discussion of topics at those meetings, the window for the conference schedule should not be coordinated more than 1 month or 6 weeks in advance (Mulders & Zender, 2021).

Case Study: Facebook's Oculus

Facebook's Oculus is a VR technology that is paramount to the metaverse today. In 2014, Facebook purchased Oculus, a US company that markets VR devices and services, for US$ 2 billion in stock and cash. Oculus was already enjoying a solid customer base and was well respected in the gaming industry. The purchase of Oculus was a tactical move to expand the emerging VR industry from gaming into new economies of scale. This move led to Facebook's development of the VR gaming platform Facebook Horizon Worlds and Facebook Horizon Workrooms, a digital remote working application. Both can be accessed with Oculus Quest devices (Kraus et al., 2022).

Oculus VR technology and content platform services and products help users access a fully immersive and interactive world to train, learn, explore content, and interact with others. Besides gaming, the new economies of scale that Oculus affords include media, communication channels, entertainment, education, and other domains. Oculus has tremendous potential for VR technology to reign as the next social and communications king (Kumar, 2019). Commercial, off-the-shelf VR headsets, such as Oculus Quest 2 (available at US$ 299–399) launched in October 2020, render reality into an affordable VR system for anyone who seeks them (Tin et al., 2021).

Meaningful Connections

This case study indicates the extent to which the metaverse is cast as a new frontier for the virtual environment. Oculus offers unique abilities to connect end users, with more intimacy and affectivity. This can be achieved through both the applications within Oculus itself and the Oculus' use of Facebook's other software (e.g., Facebook, Messenger). In doing so, Facebook sets up Oculus as one affording roles and actions similar to those in everyday life (Egliston & Carter, 2022). As Oculus' Head of Product Marketing, Meaghan Fitzgerald, remarks, the metaverse has the ability to give end users "meaningful connections" with others (Egliston & Carter, 2022, pp. 77–8). Another important Oculus executive, Michael Abrash (the Chief Scientist), depicts Oculus as a platform giving the power to create "high bandwidth" computing, as in allowing us to "do more of what makes us human, especially socially" (Egliston & Carter, 2022, p. 78). The "meaningful connections" between users to whom Fitzgerald refer are not only possible for end users. Oculus is for all content creators: all who want to share and distribute content. As a number of research and development specialists, especially those studying business models and the political economy of platforms, contend, the success of most big-tech companies depends on the facilitation of direct connections between end users and content creators. Platform owners can also benefit from these. For example, they can earn money through a stream of content that is outsourced to content creators or through a swarm of users attracted to this content (Srnicek, 2016).

Ford Motor Company

Ford Motor Company is fascinated by the potential that Oculus can offer. The US automobile producer Mogul has invested in the VR technology on the assumption that it will become more developed that its current use as an entertainment vehicle. It hopes Oculus will assist technicians and other professional users in a manufacturing capacity. Ford is thrilled because Facebook's platform adds a more user-friendly approach for countless developers to create all sorts of applications. For example, its headset is less heavy and more portable as compared to what Ford had used for several years. Ford Motor Company was already at the forefront in adapting VR technologies at the end of the twentieth century (Kim, 2016). In addition to providing a logistical edge, Oculus technologies are set up in ways that can easily enhance insight and decision-making for those who use them, as in the ability to seize and comprehend a range of physiological data about members of the workforce based on their participation in VR simulations (Egliston & Carter, 2021a).

Surveillance and Data Collection

It is important to mention that Oculus grants workplaces more leeway for surveillance (Zuboff, 2019). The implementation of contiguous identity—in which participants interact in virtual environments, but still remain bound to their physical selves—can probably become a predominant source of data collection. Oculus headsets can gather highly detailed data about users' corporeal movements and the scenes in which these movements occur. In turn, the bodily actions of VR users are associated with other profiles within the metaverse. This contiguity of performance is easier to surveil because the users' immersive, parallel experiences in VR are part of a larger data profile created about employees moving across platforms. Put simply, for each worker, data sources are combined from multiple platforms; the VR experience is not to be examined in isolation or distinct from other identity performances (Saker & Frith, 2022).

In point of fact, the data produced through VR is increasingly being used to keep an eye on workplace performance. Thus, Oculus can be framed as a form of enterprise technology—devised as a method for collaboration, simulation, enhanced efficiency, data collection, and surveillance (Egliston & Carter, 2021b). More recently, illustrating the data-borne power of Oculus VR in such a context, the employment of Oculus hardware for workplace training and hiring is something to behold, particularly in the service industry—such as retail. In 2018, Oculus announced a partnership between Walmart, the US-based multinational hypermarket chain, and

STRIVR, the leader in Immersive Learning solutions based in Palo Alto, CA, for workplace training and evaluation. Initially, a corporation that churned out VR software to train athletes, STRIVR, is today one of the largest VR EdTech and enterprise training firms in the world. It claims to have hosted more than six million training sessions, mostly through its partnership with Walmart. STRIVR professedly gives Walmart the opportunity to simulate events that would be harder to imitate in three-dimensional training settings (e.g., a Black Friday shopping crowd). The EdTech firm also teaches how to use new VR technologies before their installation, and they are a leader in soft skills training such as customer service and empathy in the workplace (Egliston & Carter, 2021a).

Workplace Employee Avatars

Like their counterparts in the general metaverse, workplace employee avatars can be highly immersive, provide sensing and data visualization devices, and display spatial analytics in the metaverse economy—this is good news for optimizing performance parameters. Workplace employee avatars are also ideal for virtual work meetings and collaboration tools (which can be improved through colossal computational resources, blockchain networks, and digital testing capabilities) (Bratu & Sabău, 2022; Lukava et al., 2022). Let us not forget the benefits of such avatars for virtual recruiting, onboarding tools, and advanced workflows because they help configure metaverse engagement metrics (Gibbert et al., 2023; Kozinets, 2023). Brands and companies have successfully used avatars in hiring workflows, distance learning, skills training, and networked practices. Workplace employee avatars are good for sensory immersion because they can forge emotional connections and give form to digital personas across virtual environments. They can do so by including haptic gloves and bodysuits, natural language processing programs, and cognitive computing systems (Ljungholm, 2022).

Integration and Participation of Employees

Workplace employee avatars can be infused within 3D networked practices for purposes of knowledge construction in groups, a process often called co-creation. That knowledge can then be evaluated by connected monitoring algorithms and movement tracking sensors. The evaluation itself is displayed through cognitive and affective metrics. Besides knowledge construction and evaluation, workplace employee avatars can enhance employee development by integrating quality tracking software, spatial intelligence devices, computer vision tools, and autonomous devices. Taken as a whole, virtual personas in highly immersive extended reality environments improve all types of collaboration and production by capitalizing on augmented reality tools and spatial data mining (Popescu et al., 2022). At the workplace, many different players can use avatars. With sufficient training (and authorization) to use them, they can help establish platforms and data exchanges for their respective organizations. Among the eight billion inhabitants today, the tech-savvy generation is undoubtedly using English as the main work language. To be certain, English remains the main lingua franca in this new reality of the metaverse. Third-party recruitment and involvement will remain key to this digital transformative journey, particularly since most of the big-tech companies that have invested billions in the metaverse hail from the Anglo-Saxon world (Allam et al., 2022).

Perception of Easiness

An employee represented by a true-to-life avatar is easier to identify, welcome by colleagues, and deemed more reliable—despite the fact that privacy is diminished with life-like avatars because they "expose" these employees (Jamadi et al., 2022). In a study conducted by Panda et al. (2022), participants felt higher social presence in a joint call when they saw the workplace employee avatar than when they did not (as would be in a traditional teleconference call). This is particularly the case in a mixed-modality conferencing model such as "Together Mode," where call participants can be seen in person, as if they were together in a virtual shared environment, building strong unity and coherence for visually present participants. Conversely, audio participants said that it was more difficult to take part in a conversation without visual representation. They also found it easier to contribute ideas and information in both avatar and virtual shared conditions.

In the same study by Panda et al. (2022), participants commented that it was easy to become oblivious to the fact that other audio-only participants were present in the call. This is a serious problem; feeling less present as a result of a lack of visual representation could cause audio participants to not participate in future calls as much. From the observer's standpoint, not only are audio participants not present in the visual field, but they are also less willing to actively converse as they would through visual representations. To give more details, participants indicated that expressing themselves through head and limb movements (even without facial expressions) was sufficient for communicating their presence in meetings; unlike fully immersive avatars, they could keep their privacy. The use of uncomplicated gestures such as nodding was already a profound social signal that helped all participants believe that they felt included and valued in the conversation.

Pros and Cons of Privacy Issues

While the authors of this book are not putting forward the idea that communication in the metaverse is automatically better or easier than daily face-to-face interactions, it is important to observe that working in the same physical space tends to have more drawbacks. Examples include noisy environments, the risk of contracting microbes or viruses, and loss of privacy. With respect to privacy, having an awareness of our coworkers has pros and cons. Nevertheless, their actions are processed and monitored through computer systems; this is why they should not give up more life details than they choose to. The WYSIWIS paradigm (What You See Is What I See) for computer-supported cooperative work usually operates on the following principle: coworkers find themselves in a tight circle of interaction where every participant gets to see the topic(s) of discourse and, into the bargain, can respond to each other's comments. This is based on the traditional workplace format where a person speaks in front of a whiteboard: here, the audience consists of a group of coworkers who develop a design together. And just like traditional workplace models, coworkers in the metaverse may leave and enter the space momentarily or may simply be thinking of other topics. Likewise, the work may be set up in such a manner as to not necessarily entail the synchronous task with identical data (Jää-Aro & Snowdon, 2001).

Digital Twins

Digital twins are massive-scale and high-fidelity digital templates and entities carbon-copied in virtual worlds. They mirror the assets and features of their physical counterparts, including the object movements, temperature, and applications. The association between the virtual and physical twins is captured by their data (Mohammadi & Taylor, 2017). Digital twins present the advantage of concretizing virtual objects such as buildings, factories, and urban works in the metaverse. They also enable the three-dimensional world and the metaverse to connect and merge around sensing data in real time. For this reason, the preponderance of organizations and institutions are developing standards for competent digital twins. Manufacturing framework and reference architectures, for example, have been planned in these standards (Kang et al., 2023).

Best Solution for Products and Services?

Digital twins are generally implemented through computer programs that produce digital replicas of physical objects, operations, or services in the metaverse. A key objective here is to facilitate the collection of data in order to create optimal simulations. These simulations will help organizations model, test, and predict what results a brand, process, or service will yield in the real world. Thanks to digital twins, corporations and institutions will also have the ability to tackle "what-if" issues in the application and discipline of more complex phenomena, for which Bibri (2023) offers a descriptive account. Digital twins have been shown to bestow customers and users with the best solutions to certain needs, as well as the ideal prompt, real-time visual representations of development schemes or interventions. Ultimately, this allows corporations and institutions to significantly reduce costs of testing and modeling, which can be considerably expensive, labor-intensive, and time-consuming in the real world (Allam et al., 2022).

Three Stages of Development

Digital twins help companies visualize metaverse operations through three stages of development. As a first step, digital twins begin in places where our physical world is digitized and are capable of periodically reflecting transformations to their virtual counterparts. More precisely, digital twins produce digital replica of physical artifacts or services so that employee avatars, as digital natives, can develop new creations in the metaverse. As a second step, digital copies progressively interface with a massive landscape. Although virtual environments frequently suffer, initially, from restricted connectivity with each other and with the physical world—a phenomenon known as "information silo," whereby an information management system cannot freely connect with other information management systems (Bainbridge, 2019)—the digitized physical and virtual worlds soon consolidate, forming the third and final step of the coexistence of physical and virtual reality. This merging occasions unequaled demands of perpetual and 3D virtual activities in the metaverse (Paananen et al., 2023). Put simply, the digitized physical and virtual entities are now merged through data (Far & Rad, 2022). The frameworks and conditions for the success of digital twins can be achieved through other technologies—e.g., sensing technologies—to preserve the same state or appearance as their corresponding products or services. The frameworks and conditions in the metaverse can even be "sent back" to the physical devices for any modification (cited in Mitra, 2023, p. 66).

After digital twins connect with the metaverse, digital natives can use devices such as smart glasses, head-mounted displays and helmets, and brain–computer interfaces to feel the life-like functions of sensory organs or anything mimicked by digital signals. The latter can also be directly emitted to the human brain, signifying that participants can feel a meta-universe state without actually getting authentic stimuli from inside and outside the body. Overall, the metaverse gains a great deal from digital twins because they can sublimate the natural and humanistic elements of the real world into the virtual world—i.e., through a process of mirror reproduction and close arrangement. For digital natives, the human brain records and produces digital signals, which allows these participants to go beyond the limitations imposed by time, space, and individual characteristics. To boot, they can experience a wide spectrum of stimulations, such as sight, hearing, touch, smell, and taste (Pu & Wang, 2022).

Workplace Training: General Perspectives

The workplace environment is quickly changing because of the metaverse. It is undeniable that training is one of the main catalysts of productivity in the workplace. VR technologies are increasingly applied to all sorts of workplace training contexts. In this respect, the metaverse stimulates organizational activities by allowing learners to use top-notch equipment for hands-on instruction—whether or not it necessitates physical presence, as would be the case for surgical operations, airplane maintenance, or manufacturing. The immersive training afforded by the metaverse is a tremendous asset to training programs, from which institutions of higher learning can benefit in their training centers. In many situations, the metaverse is set up to train staff members in soft skills that are more complex to acquire and assess, as would be the case for communicative skills such as empathy, listening, conflict management, and leadership. Metaverse-based workplace training immerses learners in real-world scenarios and offers them sound approaches for enhancing soft skills such as interpersonal relationships with peers and supervisors (Koohang et al., 2023).

Millennial and Generation Z Employees

Practitioners express concerns that a high proportion of young employees (from their late teens to their late 30 s) oppose traditional training and, over time, amass skills gaps that affect workplace outcomes (Bujang et al., 2021). Corporations rank IT talent, lack of skills, and retention as their three most pressing concerns that stifle competition (Kappelman et al., 2019). Employees are requiring professional development training (cited in Tanner & Preiksaitis, 2023, p. 3); associations between training and retention and job satisfaction have been empirically ascertained (Miller et al., 2018). Simulations and online coaching/instruction have been integrated within IT training since the 1990s (Roberts, 2018). Simulations are more open to mistakes because they generally happen without impacting the organization's outputs or data (Harvey & Mejias, 2022). Other conventional training approaches include peer-to-peer coaching, apprenticeships, and certification programs, which can be time-consuming, costly, and too dependent on external access for potential learners (Tanner & Preiksaitis, 2023).

In hands-on courses, it is not always feasible for young learners to acquire e-business experience during their business training because the associated costs are too substantial. In the metaverse, however, learners are given the chance to create situations or environments to make better and lower-cost decisions. They can gather information and build virtual workrooms to share and plan with others or be guided by skilled entrepreneurs or avatars (e.g., non-player characters and other types of online personas). The production of business ideas and the associated funding support can also be better concretized in the metaverse. By the same token, at the end of the process, learners can further interact with other users in VR who are associated with similar businesses or brands; the interactions can pertain to diverse issues (Mitra, 2023).

Metaverse Problem-Based Learning

Problem-based learning is the process by which employees are presented with open-ended problems for solutions, whether individually or in groups. It helps gain experience in problem-solving, research abilities, virtual teamwork, self-teaching, and communicative skills for the learner. In contrast to project-based learning for which the learning goals are already established from day one, the goals and outcomes of problem-based learning are open to change and suggestions and, therefore, can be jointly set with the instructors/coaches. As such, trainees can advance ingenious ideas to solve social problems and develop business plans (Mitra, 2023). This process occasionally takes place in a virtual breakout room—a virtual space that looks like a small meeting room or a separate one in the metaverse. In the virtual breakout room, a small circle of trainees can talk about a particular topic before returning to the main room. These meeting spaces are helpful because they can smooth the progress of collaborative learning and communication, peer-to-peer support, and breaks (Chandler, 2016).

This collaborative approach is not only group-based, but it also calls for trainees to solve problems, complete tasks, and become well-versed in new terminology/concepts in pairs, in groups, or through workplace employee avatars. It opens venues for learners to synergically try to address a given problem or task, which, in due course, will allow them to learn from and grow on each other. More importantly, it helps acquire critical thinking skills, just like in the real world or individual learning in higher education. In the metaverse, though, there is more information sharing and knowledge gathering because of the rapid and voluminous access to all sorts of datasets. The same can be said about problem solving (it is faster), employee engagement (it is higher), and understanding of workplace processes (it is better). Metaverse-based collaborative learning has borne its fruits in the workplace. Experienced employees successfully provide hands-on training to new employees or trainees with some level of experience with collaborative tools. Lastly, it helps hone in leadership skills, employee commitment and participation, and team spirit (Mitra, 2023).

Industrial Workplace Learning

A major application domain of the metaverse is one that consists of training staff members and implementing advanced workplace learning in the manufacturing industry (Anthes et al., 2016), thanks to gadgets such as head-mounted devices.

Practitioners, such as engineers and operators, can train (would-be) employees by having them apply these gadgets for collaboration, problem-solving, and development of new skills (Abidi et al., 2019). The pedagogical purpose for such industrial workplace learning lies fundamentally in the various learning platforms that are available at virtually no cost (Shehri, 2012). Greene (2022) also confirms that industrial specialists are increasingly resorting to using wearable close range (CR) for workplace training, often remarking that hands-on, live VR learning operates as an in situ instructional aid—i.e., as if the learners were directly situated in the physical workplace itself. Eventually, this ultramodern approach enhances both the competent acquisition of worker knowledge and the ability to better engage in interpersonal interactions with (future) coworkers (Sorko & Brunnhofer, 2019).

Industry 4.0

The notion of digital industrial learning is gaining a lot of ground as a result of its immense benefits. It is considered as one of the pillars or components of Industry 4.0. With the rapid progress in technology, digital industrial learning is now a reality more than merely a theoretical framework. That reality has been dubbed "Industry 4.0." It has been fruitfully applied to manifold manufacturing domains, including design, prototyping, and assembly training. The whole apparatus of digital industrial learning is gigantic, so the metaverse has value in many places there. It can be applied to various phases of manufacturing, although planning and conducting assembly operations comprise the biggest chunk of the manufacturing of most products and services. Still today, dealing with assembly operations during the early stages of product design remains a difficult task. As assembly is a keystone in the manufacturing process, companies should provide as much Industry 4.0 training to their employees as possible (Abidi et al., 2019).

Positive Characteristics

The positive characteristics of industrial workplace learning in the metaverse include authentic interactivity, flexibility, wider availability, and safe, customizable, low-cost environments (Haj-Bolouri & Winman, 2020).

- *Authentic interactivity*: Authentic workplace learning experiences (e.g., training, teamwork) can be easily recreated in the metaverse; for instance, an apprentice mechanic assembling virtual automobile parts can spot mistakes when parts are poorly assembled or something is squeaking. With digital industrial learning, he or she can be guided virtually to the correct the assembly problems and get feedback as an explanation (Turner et al., 2016).

- *Flexibility*: An immersive VR experience is flexible, self-paced, and quickly altered to fit the needs of the individual learner; in another example, an engineer might require VR applications that draw his or her attention to fix a mechanical glitch, whereas an operator can also be aided by VR for maintenance issues (Choi et al., 2015).
- *Wider availability*: As VR expands the availability for workplace training at the workplace, a high proportion of learners can participate in the same VR space; an illustration of this is the Cybersphere application for manufacturing industries (Fernandes et al., 2003).
- *Safe, customizable, low-cost environments*: Immersive learning systems for manufacturing industries deliver features for a controlled, simulated setting in which the learner can perform tasks in a safe manner; for instance, heightened safety awareness within a workforce can be effortlessly achieved through VR applications. The same can be said about multilingual functions and a spectrum of training scenarios. To be more precise, those immersive learning systems can safely simulate hazardous scenarios so that trainees can better identify problem areas and adjust their learning accordingly (Fernandes et al., 2003).

These positive characteristics promote industrial workplace learning in the metaverse through direct interface and socially supported applications, which is so needed in this post-pandemic era. This approach also allows for more spontaneous connections, supportive virtual mentoring, and inclusivity as it knows no barriers in regards to the types of trainees/learners. It is a dream-come-true in the sense that it can foster a culture of lifelong learning in the workplace (Zajac et al., 2022).

Workplace Training: Impact on Organizations

Metaverse-based training promotes experiential learning with clear benefits. It is a training system dealing with practices to integrate themselves within metaverse environments. In particular, learners not only undergo more interactive and flexible training, but they also maintain a level of high-quality educational content for future training plans and interactive workflows (Yilmaz et al., 2023). VR workplace learning also gives participants the opportunity to construct the cognitive links within learning environments. It is an environmental training system set up in a way that they rapidly associate the different conditions. In a specific case study of an industry, trainees quickly learned about straightaways to tight and complex curves, as well as the psychomotor skills required to understand operations fully. The point made here is that the impact that a successful metaverse training program has on organizations is evident. In due course, it is expected to have clear positive outcomes such as increased sales, higher customer satisfaction, better market shares, fewer operation costs and errors, and decreased downtime. Danylec et al. (2022) looked at the metrics attributed to the business impact of metaverse-based training.

These metrics are key performance indicators of the status and longevity of organizations. The metrics are listed as follows:

- *Lower operation costs*: Efficient technical competency training through the metaverse, when applied to and integrated into processes, results in noteworthy improvements in system efficiency. This, in turn, gives rise to a colossal decrease in costs. Cost reduction is deemed accomplished when the following is accomplished: (1) a decrease in waste material, (2) better equipment handling and maintenance, (3) enhanced processes and procedures, (4) good inventory, and (5) heightened production and manufacturing time—among others. There is a link between metaverse-based training and the accomplishment of these goals. Therefore, it is logical to determine the health of an organization by looking at the impact of the metaverse on workplace training programs.
- *Improved quality*: Improved quality is an essential attribute that determines the business impact through an effective training program because it provokes fewer errors and improves workplace performance overall. Better tasks and operations produce more ideal outcomes. Improved quality, in turn, leads to higher customer satisfaction. Again, it is cardinal to consider the value that metaverse-based training has on organizations.
- *Improved efficiency*: Improved efficiency seeks to establish the processes, procedures, and productivity through efficient training of staff and employees. Efficiency is reached when the following are reached: (1) decreased turnaround times, (2) less waste, (3) lower maintenance downtime, (4) better safety outcomes, and (5) fewer accidents—among others.
- *Higher revenue*: On the whole, the three attributes above lead to higher revenue for an organization because they help it grow its earnings or rise its profit margins. The existential goal of a firm or corporation is often identified as increasing revenues. Hence, its dedication to effective training should be to eventually make processes, procedures, and products as smooth and costless as possible. This explains why metaverse-based training remains the most sought-after model by organizations today. Furthermore, the investment in such VR approach is measurable and quantifiable—i.e., by determining the return-on-investment (ROI) to guarantee that the outcomes that emanate from the metaverse program justify the cost involved.

What these metrics demonstrate is that the metaverse can improve workplace efficiency by enhancing key business indicators and, by the same token, fostering a happier workplace. This is a basic reason why many multinational companies such as Ford—along with Nvidia, Apple, and Epic Games—have embraced this digital system (Raj et al., 2023).

References

Abidi, M. H., Al-Ahmari, A., Ahmad, A., Ameen, W., & Alkhalefah, H. (2019). Assessment of virtual reality-based manufacturing assembly training system. *The International Journal of Advanced Manufacturing Technology, 105*(9), 3743–3759. https://doi.org/10.1007/s00170-019-03801-3

Allam, Z., Sharifi, A., Bibri, S. E., Jones, D. S., & Krogstie, J. (2022). The metaverse as a virtual form of smart cities: Opportunities and challenges for environmental, economic, and social sustainability in urban futures. *Smart Cities, 5*(2), 771–801. https://doi.org/10.3390/smartcities5030040

Anthes, C., García-Hernández, R. J., Wiedemann, M., & Kranzlmüller, D. (2016). State of the art of virtual reality technology. *2016 IEEE Aerospace Conference*, 1–19. https://doi.org/10.1109/AERO.2016.7500674

Bainbridge, D. (2019). Redefining digital library boundaries: Being more than an information silo. *Journal of Information and Knowledge, 56*(6), 274–287. https://doi.org/10.17821/srels/2019/v56i6/149766

Ball, C., Huang, K.-T., & Francis, J. (2021). Virtual reality adoption during the COVID-19 pandemic: A uses and gratifications perspective. *Telematics and Informatics, 65*(5), 101728. https://doi.org/10.1016/j.tele.2021.101728

Baptista, J., Stein, M.-K., Klein, S., Watson-Manheim, M. B., & Lee, J. (2020). Digital work and organisational transformation: Emergent digital/human work configurations in modern organisations. *Journal of Strategic Information Systems, 29*(2), 1–10. https://doi.org/10.1016/j.jsis.2020.101618

Bibri, S. E. (2023). Data-driven smart eco-cities of the future: An empirically informed integrated model for strategic sustainable urban development. *World Futures, 79*(7), 703–746. https://doi.org/10.1080/02604027.2021.1969877

Bratu, S., & Sabău, R. I. (2022). Digital commerce in the immersive metaverse environment: Cognitive analytics management, real-time purchasing data, and seamless connected shopping experiences. *Linguistic and Philosophical Investigations, 21*, 170–186.

Brucks, M. S., & Levav, J. (2022). Virtual communication curbs creative idea generation. *Nature, 605*(1), 108–116. https://doi.org/10.1038/s41586-022-04643-y

Bujang, N., Paramsivam, M., Dasril, Y., & Basar, A. (2021). Training method transformation for millennials employees. *Research in Management of Technology and Business, 2*(1), 1473–1483.

Caragnano, R. (2023). Labour law: New workplaces in the metaverse and opportunities for cultural and heritage professions. *Athens Journal of Law, 9*(2), 211–228.

CBS News. (2021, August 19). Facebook launches 'horizon workrooms' for virtual offices: Here's how it works. *CBS News*. Retrieved on May 12, 2023 from https://www.cbsnews.com/news/facebook-launches-horizon-workrooms-virtual-reality-app/

Chandler, K. (2016). Using breakout rooms in synchronous online tutorials. *Journal of Perspectives in Applied Academic Practice, 4*(3), 16–23. https://doi.org/10.14297/jpaap.v4i3.216

Choi, S. S., Jung, K., & Do Noh, S. (2015). Virtual reality applications in manufacturing industries: Past research, present findings, and future directions. *Concurrent Engineering: Research and Applications, 23*(1), 40–63. https://doi.org/10.1177/1063293X14568814

Danylec, A., Shahabadkar, K., Dia, H., & Kulkarni, A. (2022). Cognitive implementation of metaverse embedded learning and training framework for drivers in rolling stock. *Machines, 10*(10), 926. https://doi.org/10.3390/machines10100926

Egliston, B., & Carter, M. (2021a). Critical questions for Facebook's virtual reality: Data, power and the metaverse. *Internet Policy Review, 10*(4), 1–23. https://doi.org/10.14763/2021.4.1610

Egliston, B., & Carter, M. (2021b). Examining visions of surveillance in Oculus' data and privacy policies, 2014–2020. *Media International Australia, 188*(1), 1329878X2110416. https://doi.org/10.1177/1329878X211041670

Egliston, B., & Carter, M. (2022). Oculus imaginaries: The promises and perils of Facebook's virtual reality. *New Media & Society, 24*(1), 70–89. https://doi.org/10.1177/1461444820960411

Far, S. B., & Rad, A. I. (2022). Applying digital twins in metaverse: User interface, security and privacy challenges. *Journal of Metaverse, 2*(1), 8–16. https://doi.org/10.48550/arXiv.2204.11343

Fernandes, K. J., Raja, V. H., & Eyre, J. (2003). Immersive learning system for manufacturing industries. *Computers in Industry, 51*(1), 31–40. https://doi.org/10.1016/S0166-3615(03)00027-7

Gibbert, M., de Groote, J. K., Hoegl, M., & Mendini, M. (2023). Recognizing new complementarities before they become common sense: The role of similarity recognition. *Organizational Dynamics, 52*(1), 100915. https://doi.org/10.1016/j.orgdyn.2022.100915

Global News Wire. (2022, January 8). Metaverse market size to worth around USD 1,607.12 Bn by 2030. *Global News Wired.* Retrieved on April 6, 2023 from https://www.globenewswire.com/news-release/2022/06/08/2458533/0/en/Metaverse-Market-Size-to-Worth-Around-USD-1-607-12-Bn-by-2030.html

Goldman Sachs. (2022, January 11). Understanding the metaverse and web 3.0. *Goldman Sachs*. Retrieved on April 6, 2023 from https://www.goldmansachs.com/insights/podcasts/episodes/01-11-2022-eric-sheridan.html

Greene, J. (2022). Ethical design approaches for workplace augmented reality. *Communication Design Quarterly, 10*(4), 16–26. https://doi.org/10.1145/3531210.3531212

Haj-Bolouri, A., & Winman, T. (2020). *Meta-requirements for immersive collaborative spaces in industrial workplace learning: Towards a design theory*. Retrieved on May 14, 2023 from file:///C:/Users/Admin/Downloads/Haj-Bolourietal2020.pdf

Harvey, M., & Mejias, R. (2022). Addressing the United States IT manpower shortage: The role of inpatriates and technical training. *Journal of Information Technology Management, 13*(3), 1–14.

Hennig-Thurau, T., Groth, M., Paul, M., & Gremler, D. D. (2006). Are all smiles created equal? How emotional contagion and emotional labor affect service relationships. *Journal of Marketing, 70*(3), 58–73. https://doi.org/10.1509/jmkg.70.3.58

Hennig-Thurau, T., Aliman, D. N., Herting, A. M., Cziehso, G. P., Kübler, R. V., & Linder, M. (2022). *The value of real-time multisensory social interactions in the virtual-reality metaverse: Framework, empirical probes, and research roadmap*. Marketing Center Münster.

Hogreve, J., Iseke, A., Derfuss, K., & Eller, T. (2017). The service–profit chain: a meta-analytic test of a comprehensive theoretical framework. *Journal of Marketing, 81*(3), 41–61. https://doi.org/10.1509/jm.15.0395

Hwang, I., Shim, H., & Lee, W. J. (2022). Do an organization's digital transformation and employees' digital competence catalyze the use of telepresence? *Sustainability, 14*(14), 8604. https://doi.org/10.3390/su14148604

Im, S., & Workman, J. P. (2004). Market orientation, creativity, and new product performance in high-technology firms. *Journal of Marketing, 68*(2), 114–132. https://doi.org/10.1509/jmkg.68.2.114.27788

Jää-Aro, K.-M., & Snowdon, D. (2001). How not to be objective. In E. F. Churchill, D. N. Snowdon, & A. J. Munro (Eds.), *Collaborative virtual environments: Digital places and spaces for interaction* (pp. 143–159). Springer.

Jamadi, B. P., Widiartanto, W., & Dewi, R. S. (2022). Identification of communication trends in business practices efforts to increase productivity and smooth communication in the workplace: Review what evidences say. *Enrichment: Journal of Management, 12*(4), 2467–2478. https://doi.org/10.35335/enrichment.v12i4.677

Kang, G., Koo, J., & Kim, Y.-G. (2023). Security and privacy requirements for the metaverse: A metaverse applications perspective. *IEEE Communications Magazine, 99*, 1–7. https://doi.org/10.1109/MCOM.014.2200620

Kappelman, L., Torres, R., McLean, E., Maurer, C., Johnson, V., & Kim, K. (2019). The 2018 SIM IT trends study. *MIS Quarterly Executive, 18*(1), 51–84. https://doi.org/10.17705/2msqe.00008

Kianto, A., Shujahat, M., Hussain, S., Nawaz, F., & Ali, M. (2019). The impact of knowledge management on knowledge worker productivity. *Baltic Journal of Management, 14*(2), 178–197. https://doi.org/10.1108/BJM-12-2017-0404

Kim, K. (2016). Is virtual reality (VR) becoming an effective application for the market opportunity in health care, manufacturing, and entertainment industry? *European Scientific Journal, 12*(9), 14–22. https://doi.org/10.19044/esj.2016.v12n9p14

Kim, M. J., Lee, C.-K., & Preis, M. W. (2020). The impact of innovation and gratification on authentic experience, subjective well-being, and behavioral intention in tourism virtual reality: The moderating role of technology readiness. *Telematics and Informatics, 49*(Issue C), 101349. https://doi.org/10.1016/j.tele.2020.101349

Koohang, A., Nord, J. H., Ooi, K.-B., Tan, G. W.-H., Al-Emran, M., Aw, E. C.-X., Baabdullah, A. M., Buhalis, D., Cham, T.-H., Dennis, C., Dutot, V., Dwivedi, Y. K., Hughes, L., Mogaji, E., Pandey, N., Phau, I., Raman, R., Sharma, A., Sigala, M., Ueno, A., & Wong, L.-W. (2023). Shaping the metaverse into reality: A holistic multidisciplinary understanding of opportunities, challenges, and avenues for future investigation. *Journal of Computer Information Systems, 63*(3), 735–765. https://doi.org/10.1080/08874417.2023.2165197

Kozinets, R. V. (2023). Immersive netnography: A novel method for service experience research in virtual reality, augmented reality and metaverse contexts. *Journal of Service Management, 34*(1), 100–125. https://doi.org/10.1108/JOSM-12-2021-0481

Kraus, S., Kanbach, D. K., Krysta, P. M., Steinhoff, M. M., & Tomini, N. (2022). Facebook and the creation of the mmetaverse: Radical business model innovation or incremental transformation? *International Journal of Entrepreneurial Behavior & Research, 28*(9), 52–77. https://doi.org/10.1108/IJEBR-12-2021-0984

Kumar, B. R. (2019). *Wealth creation in the world's largest mergers and acquisitions*. Springer.

Lee, H. J., & Hwang, Y. (2022). Technology-enhanced education through VR-making and metaverse-linking to foster teacher readiness and sustainable learning. *Sustainability, 14*(8), 4786. https://doi.org/10.3390/su14084786

Lee, Y., Kim, M., & Koo, J. (2016). The impact of social interaction and team member exchange on sport event volunteer management. *Sport Management Review, 19*(5), 550–562. https://doi.org/10.1016/j.smr.2016.04.005

Ljungholm, D. P. (2022). Metaverse-based 3D visual modeling, virtual reality training experiences, and wearable biological measuring devices in immersive workplaces. *Psychosociological Issues in Human Resource Management, 10*(1), 64–77. https://doi.org/10.22381/pihrm10120225

Lukava, T., Ramirez, D. Z. M., & Barbareschi, G. (2022). Two sides of the same coin: Accessibility practices and neurodivergent users' experience of extended reality. *Journal of Enabling Technologies, 16*(2), 75–90. https://doi.org/10.1108/JET-03-2022-0025

Madsen, H., & Matusitz, J. (2022). Benefits of Google technologies for organizations: Perspectives from adaptive structuration theory. *International Journal of Technology Management & Sustainable Development, 21*(1), 5–18. https://doi.org/10.1386/tmsd_00049_1

Martinez-Martinez, A., Cegarra-Navarro, J. G., Cepeda-Carrión, G., & Martelo-Landroguez, S. (2020). A collaboration climate and effectiveness of knowledge management. In M. H. Bilgin, H. Danis, E. Demir, & A. F. Aysan (Eds.), *Eurasian bsiness perspectives* (pp. 135–144). Springer.

McKinsey & Company. (2022). Marketing in the metaverse: An opportunity for innovation and experimentation. *McKinsey Quarterly, 58*(2), 10–21.

Miller, C. L., Grooms, J. C., & King, H. (2018). To infinity and beyond—Gamifying IT service-desk training: A case study. *Performance Improvement Quarterly, 31*(3), 249–268. https://doi.org/10.1002/piq.21263

Mitra, S. (2023). Metaverse: A potential virtual-physical ecosystem for innovative blended education and training. *Journal of Metaverse, 3*(1), 66–72. https://doi.org/10.57019/jmv.1168056

Mohammadi, N., & Taylor, J. E. (2017). Smart city digital twins. In *2017 IEEE Symposium Series on Computational Intelligence*: 1–5.

Mostajeran, F., Reisewitz, P., & Steinicke, F. (2022). Social facilitation and inhibition in augmented reality: Performing motor and cognitive tasks in the presence of a virtual agent. *IEEE Conference on Virtual Reality and 3D User Interfaces Abstracts and Workshops, 1*, 323–328. https://doi.org/10.1109/VRW55335.2022.00073

Mulders, M., & Zender, R. (2021). An academic conference in virtual reality? Evaluation of a SocialVR conference. *International Conference of the Immersive Learning Research Network, 7*, 20919710. https://doi.org/10.23919/iLRN52045.2021.9459319

Paananen, V., Kiarostami, M. S., Lee, L-H., Braud, T., & Hosio, S. (2023). From digital media to empathic reality: A systematic review of empathy research in extended reality environments. *arXiv*. Retrieved on October 25, 2022 from https://arxiv.org/pdf/2203.01375.pdf

Panda, P., Nicholas, M. J., Gonzalez-Franco, M., Inkpen, K., Ofek, E., Cutler, R., Hinckley, K., & Lanier, J. (2022). AllTogether: Effect of Avatars in mixed-modality conferencing environments. *2022 Symposium on Human-Computer Interaction for Work, Association for Computing Machinery, 8*, 1–10. https://doi.org/10.1145/3533406.3539658

Popescu, G. H., Ciurlău, C. F., Stan, C. I., Băcănoiu (Văduva), C., & Tănase (Veisa), A. (2022). Virtual workplaces in the metaverse: Immersive remote collaboration tools, behavioral predictive analytics, and extended reality technologies. *Psychosociological Issues in Human Resource Management, 10*(1), 21–34. https://doi.org/10.22381/pihrm10120222

Pu, Q., & Wang, X. (2022). The metaverse and its influence and transformation on human society. *Asia Pacific Academy of Science, 3*(1), 1–14. https://doi.org/10.54517/met.v3i1.1796

Raj, A., Sharma, V., Rani, S., Singh, T., Shanu, A. K., & Alkhayyat, A. (2023). Demystifying and analysing metaverse towards education 4.0. *International Conference on Innovative Practices in Technology and Management (ICIPTM), 3*. https://doi.org/10.1109/ICIPTM57143.2023.10118054

Reuters. (2022, May 2). Italy's Serie a enters the metaverse to showcase new way to watch soccer. *Reuters*. Retrieved on April 6, 2023 from https://www.reuters.com/lifestyle/sports/italys-serie-enters-metaverse-showcase-new-way-watch-soccer-2022-04-30/

Robbins, T. (2022, January 11). What communicators need to know about the metaverse and NFTs in 2022. *Burrelles*. Retrieved on January 12, 2023 from https://burrelles.com/what-communicators-need-to-know-about-the-metaverse-and-nfts-in-2022/

Roberts, B. (2018). Training via the desktop: Web-based training is helping employers maximize their training dollars-and their employees' time. *HR Magazine, 43*, 98–104.

Saker, M., & Frith, J. (2022). Contiguous identities: The virtual self in the supposed metaverse. *First Monday, 27, no. 3*. https://doi.org/10.5210/fm.v27i3.12471

Shehri, W. (2012). Work integrated learning (WIL) in virtual reality (VR). *International Journal of Computer Science Issues, 9*(5), 10–21.

Sorko, S. R., & Brunnhofer, M. (2019). Potentials of augmented reality in training. *Procedia Manufacturing, 31*, 85–90. https://doi.org/10.1016/j.promfg.2019.03.014

Srnicek, N. (2016). *Platform capitalism*. Polity Press.

Tanner, A., & Preiksaitis, M. (2023). Perspectives of learning and training practitioners on gamification. *CORALS' Journal of Applied Research, 1*(1), 1–20. https://doi.org/10.58593/cjar.v1i1.12

Tariq, S., Abuadbba, A., & Moore, K. (2023). Deepfake in the metaverse: Security implications for virtual gaming, meetings, and offices. *arXiv*. Retrieved on May 18, 2023 from https://arxiv.org/pdf/2303.14612.pdf

Tin, D., Hertelendy, A. J., & Ciottone, G. R. (2021). Disaster medicine training: The case for virtual reality. *American Journal of Emergency Medicine, 48*, 370–371. https://doi.org/10.1016/j.ajem.2021.01.085

Turner, C. J., Hutabarat, W., Oyekan, J., & Tiwari, A. (2016). Discrete event simulation and virtual reality use in industry: New opportunities and future trends. *IEEE Transactions on Human-Machine Systems, 46*(6), 882–894. https://doi.org/10.1109/THMS.2016.2596099

Yfantis, V., & Ntalianis, K. (2022). Exploring the potential adoption of metaverse in government. In I. Jeena Jacob, S. K. Shanmugam, & I. Izonin (Eds.), *Data intelligence and cognitive informatics* (pp. 815–824). Springer.

Yilmaz, M., O'Farrell, E., & Clarke, P. (2023). Examining the training and education potential of the metaverse: Results from an empirical study of next generation SAFe training. *Journal of Software: Evolution and Process, 35*(9), e2531. https://doi.org/10.1002/smr.2531

Zajac, S., Randall, J., & Holladay, C. (2022). Promoting virtual, informal learning now to thrive in a post-pandemic world. *Business and Society Review, 127*(S1), 283–298. https://doi.org/10.1111/basr.12260

Zhang, G., Cao, J., Dong, L., & Qi, J. (2022). Popularity of the metaverse: Embodied social presence theory Pperspective. *Frontiers in Psychology, 13*, 997751. https://doi.org/10.3389/fpsyg.2022.997751

Zuboff, S. (2019). *The age of surveillance capitalism: The fight for a human future at the new fontiers of power*. PublicAffairs.

Chapter 7
Education in the Metaverse

The metaverse has facilitated the use and development of immersive technologies in teaching and learning. In fact, the metaverse is increasingly being lauded for its educational value, its interactive and collaborative activities, and its ability to transmit new knowledge to learners. Affordable virtual tools and applications powered by immersive technologies are expanding new horizons that can change the ways in which we learn and interact with others. In both the Western and non-Western world, teaching and learning are being transformed by the metaverse; the educational process is morphing into an interactive experience (whether with other learners or with objects in the virtual environment) (Leung et al., 2018). This process is achievable because of immersive technologies, which have tremendous transformative potential in education, and it has already drawn the attention of a great number of educational researchers. As a result, an increasing number of scholars and practitioners alike aspire to integrate the metaverse into their respective areas of expertise. It has been widely ascertained that immersive technologies will surpass in-person learning and teaching, owing to the meteoric rise of augmented reality (AR), virtual reality (VR), mixed reality (MR), and extended reality (XR) technologies over recent years (Tan, 2023).

The educational metaverse is also beneficial because it is seen as a viable solution to complex educational problems—e.g., by reconfiguring the relationships between individuals, groups, and technologies through the amalgamation of reality and imagination. The COVID-19 pandemic has posed a challenge to the global education industry in unprecedented ways, notwithstanding the fact that online education had already been considered a contentious debate for years (Zhai et al., 2022).

Additionally, because many disciplines in institutions of higher learning are oriented to the real-world marketplace, it is reasonable to expect that most types of metaverse-driven educational programs are encouraged to adapt to workplace training as well. While some concentrate on a specific domain of science and engineering, others consider treated education as one giant metaverse application among many (Suh & Prophet, 2018). An educational metaverse encompasses academic

J. Matusitz, J. Dacas, *Communication in the Metaverse*, Signals and Communication Technology, https://doi.org/10.1007/978-3-031-63485-7_7

institutions such as middle-schools, high-schools, colleges, universities, and professional training centers, as well as corporate entities or professional spaces that grant learners the opportunity to develop real-life knowledge and professional expertise that would be less feasible in certain real-world circumstances—i.e., on account of unavailability of essential resources, high costs, or other issues. The learners can take part in various metaverse training and development programs such as hands-on laboratories and (un)paid internships, irrespective of time and space constraints. Taken as a whole, it can be a low-cost platform for professional training (Mitra, 2023).

Centennials: New Generations of Students

Also known as Generation Z, Gen Z, Zoomers, Post-Millennials, Digital Natives, Neo-Digital Natives, or even the Pluralist Generation, the Internet Generation, the iGeneration, iGen, Net Gen, and the Homeland Generation (e.g., see Horovitz, 2012), the Centennials make up the majority of new generations of students. Born between 1997 and 2012, these students have grown up in a virtual and physical reality concurrently. They have learned differently than previous generations and have developed an attention span that is even shorter than that of Millennials. Educators need to be cognizant of these changes in student populations, and advocate educational solutions that set them for success in the international arena (Hernandez-Pozas & Carreon-Flores, 2019). Khandelwal and Upadhyay (2021) assert that, because Centennial students have lived their entire life in a virtual and physical reality, they tend to be more addicted to watching movies, playing games, listening to music, placing phone calls, sharing locations, among doing other things, with one device that can be placed inside their pocket. They feel extraordinarily cozy with technology.

As Centennial students are populating our virtual and traditional classrooms, educators often comment that conventional teaching methods such as lectures are not as effective. Students report being impatient, distracted, irritated, and less likely to pay attention to educators who do not provide something that Centennials consider "hands-on" and worth their time and effort. If truth be told, teachers and professors have to contend with so many distractors that impede appropriate preparation for students in the real world. At the same time, with a reduction in students' attention span, complexity of day-to-day problems rises, in addition to certain categories of knowledge and information that are being lost. Under these circumstances, academic content and practical experience are being integrated into the educational metaverse, so students can appreciate the value of immersive courses and stay engaged in such new categories of knowledge and information. In the process, they will feel more able to master other competencies (Hernandez-Pozas & Carreon-Flores, 2019).

General Examples of Education in the Metaverse

In general, metaverse-led education programs for teachers are advanced forms of digital learning or e-learning, giving them the opportunity to teach information, create activities within the tasks considered, and apply novel ideas, actions, and experimentations. These can be complemented with traditional methods of teaching and even adapted to their work practices. Scholars and researchers favor a duality of metaverse-based and conventional education so that students get ideally prepared for new professional practices (Gonzalez-Moreno et al., 2023). Any approach that favors the stimulation and facilitation of students' learning for adapting to the marketplace (upon graduation) is key. An illustration is the virtual conversational platform, which can be based on digital personal assistants or chatbots. Such platforms include enhanced sensory channels through which the platform can detect facial expressions and emotions. Consequently, they allow learners to "feel" conversations when interacting during group activities or teamwork. This anticipated evolution, through immersive technologies, in teaching and learning fosters enormous optimism and excitement for educational researchers (Tan, 2023).

The metaverse offers other promising applications and competencies in education. Some of these include the potential to give learners an unprecedented range of experiences and aptitudes, so they can (1) increase their understanding of the virtual environment in real-time, (2) engage with 3D avatars or objects, and (3) imagine abstract concepts and the complex relationships between variables in VR. In addition to collaborating with classmates in a virtual setting, students can "visit" or interact with events that are inaccessible on account of barriers such as distance, time, cost, or safety reasons. With VR headsets, students can delve into 3D spaces and unchartered territories, or expensive places and events across the world—with just a few steps operated by clicks or finger touches. Through VR, students in the field of emergency management can learn through simulated emergencies or hurricanes, volcanic activities, and so on. Likewise, students in archaeology and history are now able to walk through ancient cities (while, in reality, they are standing in a room) (Leung et al., 2018).

Naturally, students in the metaverse can interact directly with teachers through their avatars, generating an immersive learning opportunity that boosts the students' desire for learning (Wang et al., 2022). As such, Siyaev and Jo (2021) examined the application of mixed reality teaching methods in the field of aircraft maintenance; they reported a positively engaging learning experience for students in that industry. González Crespo et al. (2013) analyzed educational VR applications in free metaverse courses that used OpenSim; they found that the dissemination of knowledge was smoother and more fruitful. Saundarajan et al. (2020) investigated the efficiency of Photomath, a digital learning app focused on teaching math through AR. Their findings revealed that applying AR in math can improve students' learning outcomes. Lastly, Park and Kim (2022) identified specific types of worlds in the educational metaverse, particularly survival, maze, multichoice, racing/jump, and escape rooms. These types of worlds are what many metaverse students enjoy—although a certain number of them were initially hesitant due to the juvenile appearance of these worlds.

From Few-to-Few to Many-to-Many

Traditional in-person education is a "few-to-few" experience in the classroom—generally between 15 and 50 students. Depending on the class size, the instructor imparts knowledge to the learners, including answers to individual questions and other items of information. In the bulk of classrooms, the instructor motivates students to cooperate with each other and, through communal experiences, integrate lessons with group interactions (e.g., in-class activities) in order to facilitate learning and improving memory and cognitive processing (Garon, 2023). These circumstances and attributes can already be experienced in the metaverse. Education in the metaverse can even be formed by the convergence of two global age phenomena: the massively multiplayer online role-playing game (MMORPG) and social media (Peckham, 2020).

Each of these two embodies many-to-many experiences. Let us begin by focusing on the case of MMORPG, where participants join forces to take on characters in a vigorous story setting. The objective is to learn the lessons intrinsic to the game and through a variety of challenges that seek to better the attributes of the player's character. Here, the educational metaverse consists of learning together through role play in a digital game. The many avenues for learning that emerge from MMORPG games—remember *Fortnite*, as discussed in Chap. 2—represent immense opportunities for learning through expansive social interaction. Since learners engage each other through VR headsets, goggles, or glasses, the social interaction can either be in large groups or small ones (almost like private discussions) (Garon, 2023).

In the case of social media, it can be considered a "many-to-many" interaction platform that allows students to post or exchange information so that anyone can view or consume it (Garon, 2023). In a study conducted by Salloum et al. (2021), 410 undergraduate students were surveyed to determine whether social media-driven digital learning was worthwhile. The results indicate that it was beneficial in the sense that the participants reported enjoyment in learning when it was through knowledge sharing and social media features that have a high perceived ease of use and perceived usefulness—i.e., particularly platforms such as Facebook, Twitter, Instagram, Snapchat, LinkedIn, and YouTube, which facilitate the exchange of videos and photos, sharing content, and interacting with others. Creating content with classmates in the metaverse is also common outside classroom projects, including "many-to-many" situations that mix major events, pop culture news stories, and others with actual material that was learned in a virtual class (Peckham, 2020).

Other Advantages of Metaverse-Based Education

Communicating through a digital persona can be advantageous in many situations. Digital learners have a higher potential to connect with people as opposed to non-embodied conversational agents (Lucas et al., 2017). In fact, learning with another avatar can turn the material into a less difficult task for students in comparison with a setting that has no digital instructor at all (Lin et al., 2020). In these cases, displaying social cues is of utmost importance; when a digital human takes on a role of a teacher, it can shape learning outcomes by displaying a warmer and more empathetic appearance (Loveys et al., 2020). On top of appearance, voice and facial expressions can enhance students' motivation and learning outcomes (Horovitz & Mayer, 2021).

In line with these contentions, students show a higher motivation when the instructor expresses happiness through an avatar. Fitton et al. (2020) demonstrated that teaching in the metaverse through an embodied conversational agent does not result in lower student performance, participation, and motivation as teaching in a traditional classroom setting. Students commented that they were amazed at the fact that the embodied conversational agent displayed life-like facial expressions, motivating students themselves to display nonverbal social cues. On a side note, when avatars are being employed as museum guides, they are perceived as having a friendly personality due to their strong nonverbal communication features because they give off positive first impressions and overall positive interactions (Sylaiou et al., 2020).

Communication Boundaries

As a universal communication space underpinned by smart technologies, the educational metaverse is groundbreaking in the domain of communication, whether it is about the tools of communication available or the direction that users can take during their learning and teaching experiences. Both digital teachers and learners can play various role at the physiological and psychological levels. For example, advances in speech synthesis, brain–computer interface, and brain–brain interface can assist students in gaining mastery over communication barriers at the physiological level. Conversely, the flexibility of the digital learners' own personality and abilities can also make up for learners' communication difficulties at the psychological level, as has been ascertained for learners with deaf–blindness or autism in special education. It has also been the case for nonverbal rural learners to express and communicate better. With respect to the tools of communication, the methods available are becoming more diversified—with voice, video, 3D images, and holographic programs all becoming useful for communication, augmenting the volume of information shared and enhancing the efficiency of communication overall (Zhai et al., 2022).

A More Three-Dimensional Way

In contrast to long-established internet education where flat information exchange is the centerpiece of learning, the metaverse expands the ability of knowledge creation and interactions to a multidimensional and real-time space with the assistance of a powerful digitization of assets and identities, which considerably improves teacher-to-student and peer-to-peer relationships. Because instructors and students in the metaverse acquire digital identities through digital mapping, which reconfigures the physical, cultural, psychological, and spiritual presence of the real subject, the educational metaverse is the ideal place for teaching and learning. Unlike conventional internet education, the educational metaverse actually offers a more three-dimensional way of interaction for such teacher-to-student and peer-to-peer relationships. The reason is that it not only cultivates interactions among real subjects but also engenders two new forms of mapping relationships between real subjects and avatars, as well as between avatars and avatars (Zhai et al., 2022).

Interaction Between Real and Digital

In regards to the interaction between the real subject and the digital subject in the educational metaverse, the high level of overlap that exists between the real world and the digital world increases the complementarity and equilibrium of the two worlds for educational content, thereby creating a new educational ecology. This category of interaction includes the "interaction between the real subject and the mirror ontology" and the "interaction between the real subject and the mirror other" (Zhai et al., 2022, pp. 2–7). The former is highly dependent on the digital twin, because the interaction hypothesizes that instructors and students transmit personal information to the mirror subject in real time. In this circumstance, the educational content in the metaverse can be made accessible by AI and simulation models, and students can make prognoses about their own teaching or learning dispositions by examining the learning performance tendencies or directions of the mirror subject (Zhai et al., 2022).

With the latter statement, "interaction between the real subject and the mirror other," there is, indeed, a crucial relationship between the mirror subjects in the educational metaverse. The social aspect of the educational metaverse is evident not only in the parallel association with the real society but also in the capacity of digital humans to learn on their own and to spontaneously develop social relationships. The social aspect also implies the ability to engage in discussions in the educational metaverse, generating advanced knowledge content and learning styles, so that new "digital twins" effortlessly become "digital natives." Thanks to AI technologies such as natural language processing and group-based intelligence models, mirror subjects are greatly simulated and their activities become as natural and spontaneous as possible (Zhai et al., 2022).

Linguistic Perspectives

XLS-R is Meta's massive program for improving communication and learning in multiple languages (Babu et al., 2021). An application such as XLS-R can help learners whose native languages are different to speak directly (in their own languages) in the educational metaverse; the person's talk will be interpreted directly. In other words, the foreign speaker will sound American to the US audience. It is also a good way to learn a common language and educate others. More generally, the decentralized structure of the educational metaverse transcends the restrictions of one-way communication of knowledge resources. It encourages two-way dissemination of information and produces a favorable symbiosis circle. In the past, traditional information technologies were advantageous in terms of education equity because they could be easily transferred to various classes of students or even to those living in the countryside. On the other hand, with the educational metaverse, the increasing digital divide raises concerns of equity in education. For instance, the digital aspects of course curricula are often crowding out time in the classroom, which causes students to lack in adequate interpersonal and social interactions. These problems can be somewhat alleviated inasmuch as the educational metaverse allows all groups of learners to achieve two-way sharing of fundamental resources and content. Ultimately, the goal is to reach a multiple-way output of educational equity (Zhai et al., 2022).

Digital Literacy

Digital literacy refers to a person's aptitude in locating, evaluating, and communicating information by searching it online or through digital media platforms. It is a blending of both technical and cognitive skills in applying communication technologies to create and share such information (Jones & Hafner, 2021). Analogous to other evolving definitions of literacy that stress the cultural and historical approaches to making meaning, digital literacy is not a substitute for conventional methods of interpreting information. Instead, it builds upon the foundational knowledge of these traditional literacies. In other words, digital literacy elongates the path toward acquiring knowledge (Jenkins, 2009). Now, it should be noted that digital literacy somewhat differs from technological literacy. Technological literacy denotes the knowledge and skills that one must have in order to perform everyday activities with technological devices. Put another way, it is an individual's ability to properly use computers in a successful manner (Downs, 2019). Technological literacy can be easily improved with training, as was described in the previous discussion on institutions of higher learning (those that are oriented to the real-world marketplace).

On Metaverse-Based Platforms

Mastering a substantial volume of digital literacy on metaverse-based platforms is increasingly becoming a requirement for teachers and students to skillfully apply knowledge in virtual and physical environments, and use interaction techniques with others (Gamelin et al., 2021). Possessing these skills can determine whether instructors and learners are able to navigate through immersive environments and appositely concentrate on learning tasks in metaverse-based applications (Wang et al., 2015). Therefore, digital literacy wields a significant impact on learners' ability to implement their daily knowledge of metaverse technologies successfully (Hasse, 2017), and it can be constantly improved over the course of the entire school training (Alvarez-Risco et al., 2022). Moreover, it has been proven that institutional support smooths the progress of digital literacy learning in better ways than, say, digital literacy applications that are self-taught (Shao & Kwon, 2021). This is why institutions have made efforts to provide both students and teachers with a good grip on the entire apparatus of information handling and overall navigation in virtual environments (Diamantidis & Chatzoglou, 2018). For instance, instructors and learners in English departments should keep abreast with virtual pedagogical tools to fulfill the needs of English courses in the metaverse. It is also primordial in order to develop the knowledge and language abilities in metaverse-based blended learning (described below) (Zhou, 2022).

Blended Learning

Blended learning is a pedagogical method that bequeaths the educational metaverse community with opportunities to enhance learning through synchronous and asynchronous delivery modes. By improving it both ways, the levels of interaction among the participants involved (i.e., students and faculty members) will improve in the process. Blended learning has become the standard around the world, allowing students to gain from more interesting and effective approaches to instruction, which can be easily suited to their needs (Hrastinski, 2019). Garrison and Kanuka (2004) define blended learning as "the thoughtful integration of classroom face-to-face learning experiences with online learning experiences." Thus, there is a general consensus that the keystone of blended learning is the combination of face-to-face and virtual instruction/learning. Before the concept became popular, the term "hybrid learning" was the norm. Nowadays, blended learning and hybrid learning are employed interchangeably (Graham, 2009). Akin to the descriptions of blended learning provided above, a hybrid learning environment refers to a mingling of face-to-face education and access to digital learning tools (Hall & Davison, 2007).

The Importance of Keeping Traditional Literacy

Whereas digital literacy has value and tremendous potential for growth, traditional literacy should remain indispensable. The rationale is that, as soon as digital literacy boils down to mere "literacy" and a learner no longer feels the need to be seriously engaged in offline identity narratives, then transmedial subjectivity may soon disappear (Davidson, 2022). Transmedial subjectivity is the ability to understand or interpret various perspectives in literacy, particularly when the audience is invited to enter a particular narrative world (Eiroa, 2019). Even with state-of-the-art interactional settings and life-like avatars, programming codes—in this case, related to language—in the metaverse are so advanced that students could begin to find traditional literacy less important. Hence, what we are up against is a generation of learners moving from meta language to object language. The means to codify reality are less relevant; rather, learners delve profoundly in the constructed social reality itself, making literacy and natural language seem obsolete. Thus, this situation reaches a point where language and speech may be used predominantly to describe the digital space, apparently and outwardly becoming a meta-language itself. The traditional classroom employs mediated natural language; it is a space where students and teachers learn the transmediality of cultural texts. As the educational metaverse becomes more and more prominent—not just virtually, but also interactionally, socially, emotionally, and psychologically—the transmediality of language decreases in importance, brought down to a single modality—the digital—for both literacy and identity construction (Davidson, 2022).

The educational metaverse will evolve within these spaces. Yet, it should simultaneously preserve traditional literacy approaches, emphasizing solid language skills and the construction of identity as an offline/online transmediality (instead of primarily an online-only construction). An appropriate *mélange* of literacy seems ideal and it can be applied without difficulty, and there would still be large groups of learners interacting with each other to build their identities. An appropriate application of digital literacy should take the contextualization of essential literacy and identity construction into account, highlighting the importance of both the digital and the physical. Both can complement each other without too much overreach; the learner can develop a self in a holistic fashion (Davidson, 2022).

Case Study: The "We Work" Alternative

For the educational metaverse, an adequate level of digital literacy is not only indispensable, but it also needs to be approached with adequate teaching and learning modalities. It could even be a modality in which students must attend in-person class meetings all day and still receive most of their instruction in a virtual manner. While the metaverse replaces the whiteboard and projector with its 3D immersive tools, students still put their laptops and digital devices down to complete other instructional and leisure-related activities at school, such as going to the study hall or to the gym and attending science labs and music recitals. Naturally, university administrators and faculty are concerned that, should the metaverse replace those said activities and locations, then the physical presence of students on college campuses would dwindle dramatically. What administrators and faculty prone on embracing the educational metaverse advocate is a reasonable use of VR technologies, such as headsets, to be used not only at home but also on college campuses. Into the bargain, because social welfare and government programs, as well as scholarships, are essential to many public institutions of learning, cities and states do not want brick-and-mortar schools to disappear (Garon, 2023).

Instead, college campuses tend to adopt a model based on a "We Work" alternative. Under this model, multiple locations are created (by state legislatures and school districts) to provide ample space for study rooms where top-notch VR equipment is free to use. In these rooms, students are not even separated by grade or subject matter. They can walk into the nearest community center, library, or study hall and enter an open VR suite. Then, they log into the course or class project for which they signed up. More importantly, decreasing or eliminating the use of those multiple locations as intermediaries would disrupt the learning process of a great many students, not only because they have been so accustomed to this approach but also because schools and universities (at least in the United States) have invested hundreds of billions of dollars. Although the point was made earlier that an overreach from the educational metaverse will have a detrimental effect on student's identity construction, socialization, and traditional literacy, VR design applications have tremendous value when used appropriately—"We Work" reflects such appropriate mingling of physical and virtual learning. It should also be emphasized that students with certain circumstances or special needs might be better suited to the "We Work" alternative for all the reasons mentioned in this paragraph (Garon, 2023).

What the "We Work" alternative demonstrates is that, as instructional and learning strategies evolve in ways commensurate with the metaverse, they need to take traditional concepts and approaches into account. This could be illustrated in (1) a blending of 2D and 3D, (2) an adaptation of 2D into 3D, and/or (3) an adaptation of 3D into 2D. The mingling of physical and virtual learning can provide more dynamic and realistic settings, including situations requiring eye/body movements and voice recognition. Difficult concepts, processes, and methods can be better addressed and understood through the "We Work" alternative. Hands-on scenarios and case studies based on real-life applications should not be provided only in a traditional fashion or a virtual way. Again, social connections, interactions, and communication at large can be best developed when participants are immersed in both circumstances with top-notch VR equipment and situations involving three-dimensional bodily expressions, language, touch, and smell. All in all, the "We Work" alternative recommends the maintenance of social interactions even if the metaverse becomes a fundamental pillar for interactions among students, as opposed to a strong substitute for interactions (Mitra, 2023).

Past Studies Demonstrating the Success of the Educational Metaverse

In this section, four scientific studies were selected because they demonstrate how successful metaverse-based higher education has been. The four studies were conducted by Merchant et al. (2014), Jensen and Konradsen (2018), Wu et al. (2020), and Tan (2023). One important conclusion that is common among all four studies is that instructors and students should continuously dedicate their time and energy to metaverse-based education in order to use or benefit from instruction effectively. Furthermore, the curriculum should go through clever transformations in order to shift from the lecturing of conceptual knowledge to a more hands-on emphasis. For these reasons, it is important that university administrators and faculty collaborate to design learning programs that maximize the pedagogical perks of metaverse-based innovation.

Merchant et al.'s (2014) Study

Their meta-analysis investigated the impact of instructional design approaches to immersive learning (for both K-12 and higher education settings equally). Contrary to simpler versions of learning based on simulation, the authors demonstrated that learners can display better learning outcomes by using VR applications such as digital games. More specifically, the number of learning sessions in that context is inversely proportional to the growth in learning outcomes. On the other hand, the study findings also revealed that too much involvement in virtual environments can be detrimental to learning outcomes—a useful parallel can be drawn with the loss of transmedial subjectivity described earlier. Simpler versions of simulation-based learning (i.e., desktop VR technologies) are more beneficial for conceptual tasks, whereas fuller immersive learning is more suitable for procedural tasks. The authors were able to draw their conclusions based on the appropriateness of specific VR instructional designs and tools and when properly applied to particular learning outcomes.

Merchant et al.'s (2014) meta-analysis also underscored the power of the educational metaverse to facilitate transfer of learning. Transfer of learning is characterized by the process of transferring skills and knowledge acquired in one context to another context. It can also be understood as a phenomenon whereby the accumulation of past knowledge and skills influences the performance of a subsequent (or even less related) activity (Mestre, 2002). With respect to immersive learning technologies, they can be applied to situations where learning experiences incorporate learners' digital fluency in novel ways: ways that stimulate participation, learning, and transferability between education and real-life situations (Dede, 2009). As such, the ability of students to solve various problems and get geared up for future learning is also a good illustration of transfer of learning (Schwartz et al., 2005).

Jensen and Konradsen's (2018) Study

In their study, a thorough and meticulous review was conducted in order to add fresh insights to the current body of knowledge about the use of head-mounted displays (HMDs) in education. The two authors selected and evaluated the quality of 21 published analyses (i.e., self-reported experimental studies) and then assessed their quality by means of a comprehensive search. The general conclusions indicate that HMDs are helpful in the acquisition of three major skills, namely (1) cognitive skills in regards to remembering and understanding spatial and visual data and clues; (2) psychomotor skills in regards to head movements, as would be the case for visual scanning and observational skills; and (3) affective skills in regards to controlling one's emotional reactions to distressing situations. In addition to developing these three major skills, HMDs do not seem to have an edge over simpler versions of simulation-based learning—as ascertained by Merchant et al. (2014) in the previous study—or face-to-face instruction. In fact, on some occasions, Jensen and Konradsen's examination warns about the detrimental effects of HMD-based learning when it is excessive because it can be counterproductive or cause cybersickness. Owing to their high immersive nature, HMDs can also distract learners from the task in question.

It is also important to note that, even though Jensen and Konradsen's study was focused on immersive VR technologies in learning, several key omissions need to be addressed. First, their study does not look at immersive VR applications based on the actual substance of learning or the subject matters. In addition, their study fails to examine immersive VR applications vis-à-vis design elements, major learning theories, and fields/disciplines to which they could be applied. Similarly, their findings are limited to the use of HMDs in domains such as conceptual knowledge, gesture skills, and emotional regulation. Notwithstanding the fact that their study stresses education, it does not tackle central issues of learning.

Wu et al.'s (2020) Study

It is also a comprehensive review of the literature on immersive learning. Here, it identified studies published between 2013 and 2019. More specifically, the authors selected 35 randomized controlled trials or quasi-experiments. The overall findings revealed that HMD-based immersive learning is more productive than non-immersive learning styles with a small-impact size. More detailed findings can be categorized as follows: HMD-based immersive learning has a greater impact than typical VR-based learning (1) on students in K-12 than university students; (2) in the domains of science education and special education (i.e., particularly specific abilities development); (3) when exposing students to advanced VR representations; and (4) in comparison with lectures or real-world phenomena. The authors' comprehensive review confirms that, when used properly and not excessively, HMDs can enhance knowledge and skill development and uphold learning effectiveness.

Tan's (2023) Study

Tan's meta-analysis of journal articles analyzed the application of immersive technologies to learning designs in a whole range of disciplines in US higher education. Overall, the study finds that, between 2011 and 2020, there was an increase in the application of VR in university-taught courses and training sessions to offer a more authentic learning experience to students. Approximately 72% of the journal articles pertained to studies involving high-end HMDs in VR-based learning. For example, HMDs such as Oculus Rift and HTC Vive were used. Many of these up-to-the-minute VR applications offered various accessories, such as touchpads, controllers, and haptic feedback suits. On the other hand, the study also pointed out that a limited number of studies included VR technology accessories that were low-budget mobile VR, as explained by Masuoka et al. (2019). Nevertheless, a clever use of low-budget devices and platforms can still yield highly positive learning outcomes for students in US higher education, regardless of the discipline. Of equal relevance is the fact that low-budget devices and platforms not only give more access to less-savvy educators and disadvantaged students, but they also shift the emphasis of metaverse-based learning on relying on top-notch technologies to the ability of simpler (but clever) ways to deliver high-quality pedagogical experiences to students.

Tan's study identified engineering as the most popular sphere of application in 24% of the articles. Other popular domains were computer science (12%) and astronomy (8%). Taking all journal articles into account, more than 30 articles could be classified in cross-multiple categories (Tan, 2023). The meta-analysis also indicates that VR applications in US higher education were often used to teach (1) procedural-practical knowledge (33%), such as filing a report (Shamsudin et al., 2018) and putting out fires (Zhang et al., 2017); (2) conceptual knowledge (25%), such as learning about corrosion prevention (Webster, 2016) and essential terms in organic chemistry (Edwards et al., 2019); and (3) analytical and problem-solving skills (12%), as would be the case in diagnosing patients (Cecil et al., 2018) and learning how to code (Dass et al., 2018). The remainder of the learning content categories were in the fields of communication, teamwork, and soft skills (10%), behavioral impact (6%), and learning a foreign language (2%). The other categories that were not included accounted for 6% and 4% (Tan, 2023).

Taken as a whole, the meta-analysis revealed that the "realistic surroundings" and "basic interaction" design elements were identified in all types of VR applications. Consequently, these design elements can be regarded as fundamental requirements for educational immersive applications. Learning outcomes—particularly factual, conceptual, and procedural knowledge, along with transfer of learning—had a high correlation with the use of deep metaverse-based education. Conceptual knowledge is predominantly imparted in lectures, where students are required to memorize what they have been taught for exams in the courses. In this manner, immersive applications are helpful for teaching conceptual knowledge and making sure that learning motivates students to participate. The design elements of the bulk of conceptual knowledge applications depend on two other elements: (1) a soft start for immersive learning development, which tends to be easy for students and instructors to use; and (2) any important curriculum changes that take place. The application of immersive learning designs that enhance students' conceptual understanding of a subject matter can be included as an introduction to immersive learning in US higher education (Tan, 2023).

Current Concerns with Distance Learning

Since the COVID-19 pandemic, one predominant component of virtual education has been distance learning. Distance learning is defined as "a form of class in which teaching-learning activities are done at different times or spaces while the instructor and learners do not meet face-to-face" (Hyun, 2021, p. 130). Distance learning was already growing before the pandemic, and the current status of its development is divided into three classifications, usually as a function of the media used for delivering content to students (Hyun, 2021):

- *E-Learning (Electronic Learning)*: an internet-based learning modality in which no restrictions (in time and space) have been placed on either the students or the instructors. Naturally, learning occurs through information and communication technologies, and a high level of interaction is still possible among all parties involved. Of equal note is that self-directed activities can be added to the supplier-oriented method (Njenga & Fourie, 2010).
- *M-Learning (Mobile Learning)*: a learning environment that employs portable wireless devices, such as personal digital assistants, portable PCs, and smartphones (Korucu & Alkan, 2011).
- *U-Learning (Ubiquitous Learning)*: a human-centered customized learning modality in which students can learn in any shape or form, at any place, and at any time (Cárdenas-Robledo & Peña-Ayala, 2018).

Based on these three classifications, it can be deduced that openness (i.e., no time or space constraints), flexibility (where students are free to determine the progress and the option of self-learning is on the table), and dispersibility (i.e., by using diverse learning resources at any time or place) are the main constituents of distance learning. As a result, learners amass a volume of knowledge and interact with others through different modalities. In order to form creative talent, it is also important to come up with solutions that are flexible and adaptable to a future society. Speaking of society, one major concern is that it is becoming a society where the Fourth Industrial Revolution—i.e., the next phase in the digitization of the manufacturing sector—continues to expand exponentially. At the same time, the evolution of distance learning platforms should be commensurate with students' and educators' abilities (and willingness) to apply the principles of distance learning in their daily tasks, not just for the sake of building a digitized future (Heonjun, 2021).

Additional concerns with distance learning are the following: upon examining a survey of "awareness, utilization, and experience of distance learning" conducted online in August 2020 with almost 3000 faculty members and almost 30,000 students at various US universities by the Ministry of Education, the top difficulties in distance learning, according to the student respondents, were "lacking communication with the instructor and other learners" (59.2%) and "decline in concentration" (54.3%), while faculty members responded that the problems principally pertained to "operating class according to the characteristics of the subject" (45.7%) and "motivating and students' learning and inducing their participation" (45.6%). Both teachers and learners expressed concerns that communication in distance learning (and the willingness thereof) should be improved (cited in Hyun, 2021, pp. 130–1).

Social Cognitive Theory (SCT)

In 1986, Social Cognitive Theory (SCT) evolved from Social Learning Theory, initially proposed by Bandura (1965). SCT postulates that learning by individuals takes place in a social context; it is a reciprocal and dynamic process of interaction between people, behaviors, and the environment. The theory also considers an individual's past experiences that influence behavioral action. It seeks to explain how humans regulate their behavior through control and reinforcement in order to perform a particular behavior intended to accomplish a goal; such behavior can be upheld over time (Bandura, 1986).

In addition, SCT explains the methods by which people can be motivated for behavioral change. It rests on the premise that people do not have to personally assume a new behavior; rather, they learn through observation. By observing others, they can also better understand the context/environment and other behaviors, which serves as a primary factor that influences the learning process. When humans observe a model enacting a behavior, they recollect the consequence of the events and use this memorized knowledge to direct their subsequent behaviors. This phenomenon is sometimes referred to as the vicarious learning process, or vicarious reinforcement. Taking heed of positive consequences of a behavior will inspire individuals to mimic the behavior, whereas taking note of negative consequences of a behavior will deter them from performing that behavior (Austad et al., 1984).

SCT in the Metaverse

SCT posits that learning occurs not merely through direct experiences, such as those in the classroom, but also through vicarious experiences, as in observation or social interactions (Bandura, 2001). This process is also true in the metaverse, where users can take on an adequate level of similarity from those that they observe (i.e., be it a real person or a digital person, such as an avatar). In the metaverse, it is possible to achieve the same anticipated results by mirroring the observed behavior. In other words, inasmuch as users feel they can absorb and replicate similar features (e.g., skills and efficacy), they will be more motivated to learn content (and show favorable responses) in virtual environments (Torres & Briggs, 2007).

Passive engagement in the metaverse alone can still lead to an intense relationship among people when observation is a crucial component. Active engagement is not a necessary first step. In SCT, each person possesses a unique collection of experiences that allows him or her to assess and regulate their thoughts, emotions, motivation, and actions. It is a self-system providing referential mechanisms that facilitate the perception, regulation, and evaluation of behaviors, forming an interplay among the interactants in the wider system. To put it in layperson's terms, the metaverse is so open that it can help individuals shape their cognitive processes by observing others (Alvarez-Risco et al., 2022).

Applying It to Learning

SCT is a learning theory predicated on the notion that individuals learn by observing others, in their environment, in terms of how they behave. Thus, cognition is a primary factor that guides development in a reciprocal relationship or in multidimensional ways. The theory concentrates on knowledge acquisition and learning processes that are directly tied to the observation of models. Put differently, when people observe a model enacting a behavior and, as a result, displays the effects of that behavior, they can easily store the sequence of events in their memory and subsequently use it as a guide to behaviors (Leung et al., 2018).

On digital platforms, students can not only learn by observing how classmates interact during live meetings or team-based projects, but they can also learn by replicating the actions of the instructors themselves. This would be the case in hands-on courses such as public speaking—among many others. It is also worth noting that students' ability to replicate modeled behavior depends on the type virtual platform available in the metaverse. According to the tenets of SCT, the setting is of utmost importance because it strongly influences the type of behavior—or the person's way of thinking (cognition)—that students can engage in. A case in point are the different types of learning. Centennials prefer visual learning (i.e., videos, images) and experiential learning (e.g., simulations, serious games) owing to their frequent use of technology. Conversely, they do not favor auditory learning (e.g., lectures) as much. Having been exposed to fast-paced multimedia their whole life, they find it harder to concentrate on long readings, elaborate logical explanations, and complex problems. The educational metaverse is designed to support all types of learning; it can be catered to all generations (Hernandez-Pozas & Carreon-Flores, 2019).

The Immersive, First-Person Perspective of VR

The immersive, first-person perspective of VR maximizes the range of possibilities for learning through observation and emulation. Evidently, even traditional classroom environments can facilitate this observation process for students. However, they are limited in time and space. With the metaverse, students from all four corners of the planet can immerse themselves and observe others when enrolled in the right course or project. Observation often occurs through situated learning because behaviors are not always prepared or anticipated; instead, they emerge under particular circumstances. This is beneficial not only for students. As such, staff training and teaching can be highly effective when they include experiential learning techniques because they focus on user-centered behaviors far more than didactic teaching and straight lectures. The immersive, first-person perspective of VR is a form of experiential learning because it involves interactive teaching that allows students to apply techniques in activities, such as roleplays. It is a 360-degree perspective of immersiveness where interactivity and experiential learning enable learners to step into other people's shoes by observing them. This, in turn, adds more empathy and compassion to the learning experience (Riches et al., 2022).

Social Agency Theory

A related theory is social agency theory. The theory postulates that individuals in society have the power to perform a range of behaviors because interactants feel that they have social agency. More specifically, social agents can use a wide spectrum of verbal and visual cues that they want or that fit the situation. This stands in sharp contrast to social structure theory (or structuralism), which rests on the assumption that people's actions are inhibited by social structures (Atkinson et al., 2005). In the metaverse, social agency theory reflects the idea that virtual learning environments can be established to encourage learners to follow the assumption that their relationships with others in the digital realm are social, in which traditional habits of peer-to-peer communication apply. In essence, the use of verbal and visual cues, especially human-like ones (as opposed to overtly artificial ones), in metaverse learning environments can inspire students to consider their digital interactions to be similar to that with which they are familiar during face-to-face conversations. The learning aspect of social agency is predicated on the assumption that desirable learning outcomes will ensue because this particular metaverse experience triggers positive emotions and will cause students to try harder in their learning and interactive processes (see Makransky & Mayer, 2022; Yang & Ryu, 2021).

A centerpiece of social agency theory is that social cues in instructional messages (e.g., digital pedagogical agents with whom students can identify) can strike a responsive chord in the learner (as the affection and sentiment that the teacher is a social partner). Another important consideration is the tendency for students to perceive the instructor not only as a social partner but also as a facilitator in diverse conversations. This arouses cognitive activity designed to get students to try harder and interpret what the instructor communicates. The final consideration is that students can embed themselves in more profound cognitive processing during learning. Frequent examples include mentally organizing the material into a coherent whole and relating it to past relevant knowledge available from long-term memory. This leads to favorable learning outcomes. In this context, social agency theory explains how students can relate to digital pedagogical agents through social responses aroused in learners. Ultimately, that causes them to provide more effort to engage with the material and, thereby, have more control of learning outcomes (Makransky et al., 2019). Schroeder et al. (2013) contend that the embodiment of the agent does not have to be human-like in order to generate social agency. In fact, it is frequently the case that a non-humanoid agent (such as an animal avatar) prompts similar social responses, such as a human agent would.

Self-Efficacy: A Foundation of SCT

Self-efficacy is foundational to social cognitive theory. Self-efficacy denotes an individual's confidence level in his or her potential to successfully carry out a specific behavior. Self-efficacy is inherent to SCT and has been linked to other theories as well. Most descriptions of self-efficacy position it as a phenomenon shaped by an individual's specific abilities, but also by other factors—individual and environmental factors. In a nutshell, self-efficacy can help predict the intent of a specific behavior (Niu et al., 2021; Pan, 2020). The idea that an action can be performed and maintained until it creates effects in an efficient manner has been characterized as a key predictor of human behavior (Bandura, 1989).

In this context, self-efficacy theory explains the degree to which a student can perform when engaging in learning activities (Chemers et al., 2001). It alludes to the effort on a mission that is pursued when the learner feels that he or she is competent in accomplishing that mission (Bandura, 1977). Self-efficacy is a useful measure when evaluating the efficiency of educational metaverse materials. It remains one of the central features of cognitive processing in learning. By identifying, organizing, and integrating new information with past knowledge, meaningful learning generally emerges (Fiorella & Mayer, 2015). The initial premise of self-efficacy—i.e., the individual's confidence level in his or her potential to successfully perform a specific behavior—also positively influences the notion of proximal development in metaverse interactions among students. To be more precise, the zone of proximal development refers to the difference between what learners can achieve today with what they are potentially capable of doing after multiple experiences in the educational metaverse (Vygotsky, 1978). This is why self-efficacy theory can form the basis for describing the educational experiences of students under these circumstances.

Social Self-efficacy

Social self-efficacy was briefly addressed in Chap. 3. It is the perceived competence that one has in one's social skills. Metaverse platforms offer users great opportunities to enhance their social skills and, thereby, their self-confidence in developing and keeping important connections with classmates and instructors in the metaverse. Thus, this is a cognitive–social learning model of education that stresses the value of social skills training, which also requires additional processes—i.e., knowing, modeling, role playing, and reinforcing (Ladd & Mize, 1983). Instructors can incorporate all kinds of services and activities in their virtual educational designs to facilitate the social skills training process, especially as the metaverse is often touted as a safer environment for learning. Such training involves engaging in teamwork, problem-solving with classmates, and developing approaches to support one another. In this fashion, an enhanced sense of social presence in the metaverse would further facilitate supportive interactions among users and the co-learning of positive social skills (Oh et al., 2023).

Enhancing Social Competence

Self-efficacy is the centerpiece of a person's competence, motivation, professional engagement, and personal achievement—at work, at school, and in life overall. The more potential a person believes in achieving the necessary social competence to succeed in a particular context, the more likely he or she will achieve the desired result after providing the adequate amount of effort and skill. It is akin to a self-fulfilling prophecy. Failure is unavoidable if the proper effort and skill is not delivered in the face of difficulties (Zimmerman, 1995). It is important to note that self-efficacy is not necessary an individual competence. In many cases, it needs social support. Supportive interaction from classmates and instructors (in addition to other relevant employees from campus) can undoubtedly enhance a student's social competence. It has been long established that humans can develop their social competence by learning from others in various social situations—even others with different points of view—and by forming new relationships, thereby accumulating experiences and background knowledge (Gecas, 1989).

Allowing opportunities for social competence among students is a chief objective of the educational metaverse. Studies (e.g., Valkenburg & Peter, 2008) have demonstrated that the higher the number of communication partners in the digital world, the higher the social competence in that realm. Because self-efficacy entails a belief in one's own ability, its massive motivator is the experience of mastery (Bandura et al., 1999). Hence, the more opportunities to hone in one's social skills, the more willing one engages in positive social interactions, in turn leading to better social competence (Hetherington et al., 1999). As mentioned a little earlier, the metaverse is a safe space for people to experience unparalleled social situations and expand their horizons (Ducheneaut & Moore, 2005).

Finally, thanks to their avatars, interactants in the metaverse can engage in various forms of verbal and nonverbal communication which, in and of themselves, make social interactions more vivid and salient than traditional asynchronous forms of online communication (Antonijevic, 2008). Due to a high sense of social presence, such avatar-based interactions can also facilitate learning by motivating students to develop social competence (Peterson, 2006). By emulating the avatars displayed in the educational metaverse, successful experiences of exchanging social support are more likely, thereby increasing students' confidence and fostering a social environment favorable to their learning experiences and well-being (Oh et al., 2023).

References

Alvarez-Risco, A., Del-Aguila-Arcentales, S., Rosen, M. A., & Yáñez, J. A. (2022). Social cognitive theory to assess the intention to participate in the Facebook Metaverse by citizens in Peru during the COVID-19 pandemic. *Journal of Open Innovation: Technology, Market, & Complexity, 8*, 1–14. https://doi.org/10.3390/joitmc8030142

Antonijevic, S. (2008). From text to gesture online: A microethnographic analysis of nonverbal communication in the second life virtual environment. *Information, Communication & Society, 11*(2), 221–238. https://doi.org/10.1080/13691180801937290

Atkinson, R. K., Mayer, R. E., & Merrill, M. M. (2005). Fostering social agency in multimedia learning: Examining the impact of an animated agent's voice. *Contemporary Educational Psychology, 30*(1), 117–139. https://doi.org/10.1016/j.cedpsych.2004.07.001

Austad, Carol S., Rollin S., Jane D., Dennis, G., & Judy, S. (1984). Studies of imitative learning and vicarious reinforcement. *Psychological Reports, 55*(1), 279–289. https://doi.org/10.2466/pr0.1984.55.1.279

Babu, A., Wang, C., Tjandra, A., Lakhotia, K., Xu, Q., Goyal, N., Singh, K., von Platen, P., Saraf, Y., Pino, J., Baevski, A., Conneau, A., & Auli, M. (2021). XLS-R: Self-supervised cross-lingual speech representation learning at scale. *arXiv*. Article 2111.09296.

Bandura, A. (1965). Influence of models' reinforcement contingencies on the acquisition of imitative responses. *Journal of Personality and Social Psychology, 1*(6), 589–595. https://doi.org/10.1037/h0022070

Bandura, A. (1977). Self-efficacy: Toward a unifying theory of behavioral change. *Psychological Review, 84*(2), 191–215. https://doi.org/10.1037/0033-295X.84.2.191

Bandura, A. (1986). *Social foundations of thought and action: A social cognitive theory*. Prentice-Hall.

Bandura, A. (1989). Human agency in social cognitive theory. *American Psychologist, 44*(9), 1175–1184. https://doi.org/10.1037/0003-066X.44.9.1175

Bandura, A. (2001). Social cognitive theory: An agentic perspective. *Annual Review of Psychology, 52*(1), 1–26. https://doi.org/10.1146/annurev.psych.52.1.1

Bandura, A., Freeman, W. H., & Lightsey, R. (1999). Self-efficacy: The exercise of control. *Journal of Cognitive Psychotherapy, 13*(2), 10–21. https://doi.org/10.1891/0889-8391.13.2.158

Cárdenas-Robledo, L. A., & Peña-Ayala, A. (2018). Ubiquitous learning: A systematic review. *Telematics and Informatics, 35*(5), 1097–1132. https://doi.org/10.1016/j.tele.2018.01.009

Cecil, J., Gupta, A., & Pirela-Cruz, M. (2018). An advanced simulator for orthopedic surgical training. *International Journal of Computer Assisted Radiology and Surgery, 13*(2), 305–319. https://doi.org/10.1007/s11548-017-1688-0

Chemers, M. M., Hu, L.-t., & Garcia, B. F. (2001). Academic self-efficacy and first year college student performance and adjustment. *Journal of Educational Psychology, 93*(1), 55–64. https://doi.org/10.1037/0022-0663.93.1.55

Dass, N., Kim, J., Ford, S., Agarwal, S., & Chau, D. H. P. (2018). Augmenting coding: Augmented reality for learning programming. *Proceedings of the Sixth International Symposium of Chinese, 6*, 156–159. https://doi.org/10.1145/3202667.3202695

Davidson, K. J. (2022). *Semiotic modelling of identity and communication in virtual reality, augmented reality, and mixed reality*. University of Tartu.

Dede, C. (2009). Immersive interfaces for engagement and learning. *Science, 323*(5910), 66–69. https://doi.org/10.1126/science.1167311

Diamantidis, A. D., & Chatzoglou, P. (2018). Factors affecting employee performance: An empirical approach. *International Journal of Productivity and Performance Management, 68*(1), 171–193. https://doi.org/10.1108/IJPPM-01-2018-0012

Downs, E. (2019). *The dark side of media and technology*. Peter Lang.

Ducheneaut, N., & Moore, R. J. (2005). More than just 'XP': Learning social skills in massively multiplayer online games. *Interactive Technology and Smart Education, 2*(2), 89–100. https://doi.org/10.1108/17415650580000035

Edwards, B. I., Bielawski, K. S., Prada, R., & Cheok, A. D. (2019). Haptic virtual reality and immersive learning for enhanced organic chemistry instruction. *Virtual Reality, 23*(4), 363–373. https://doi.org/10.1007/s10055-018-0345-4

Eiroa, B. S. (2019). *Transmedia narratives in the Netflix era: A case study of Marvel's Daredevil*. Universitat de Vic-Universitat Central de Catalunya.

Fiorella, L., & Mayer, R. E. (2015). *Learning as a generative activity: Eight learning strategies that promote understanding*. Cambridge University Press.

Fitton, I. S., Finnegan, D. J., & Proulx, M. J. (2020). Immersive virtual environments and embodied agents for e-learning applications. *PeerJ Computer Science, 6*, e315. https://doi.org/10.7717/peerj-cs.315

Gamelin, G., Chellali, A., Cheikh, S., Ricca, A., Dumas, C., & Otmane, S. (2021). Point-cloud avatars to improve spatial communication in immersive collaborative virtual environments. *Personal and Ubiquitous Computing, 25*(4), 467–484. https://doi.org/10.1007/s00779-020-01431-1

Garon, J. M. *Legal considerations for offering metaverse-based education*. Retrieved on March 3, 2023 from https://ssrn.com/abstract=4323227

Garrison, D. R., & Kanuka, H. (2004). Blended learning: Uncovering its transformative potential in higher education. *Internet and Higher Education, 7*(2), 95–105. https://doi.org/10.1016/j.iheduc.2004.02.001

Gecas, V. (1989). The social psychology of self-efficacy. *Annual Review of Sociology, 15*(1), 291–316. https://doi.org/10.1146/annurev.so.15.080189.001451

González Crespo, R., Escobar, R. F., Aguilar, L. J., Velazco, S., & Castillo Sanz, A. G. (2013). Use of ARIMA mathematical analysis to model the implementation of expert system courses by means of free software OpenSim and Sloodle platforms in virtual university campuses. *Expert Systems with Applications, 40*(18), 7381–7390. https://doi.org/10.1016/j.eswa.2013.06.054

Gonzalez-Moreno, M., Andrade-Pino, P., Monfort-Vinuesa, C., Piñas-Mesa, A., & Rincon, E. (2023). Improving humanization through metaverse-related technologies: A systematic review. *Electronics, 12*, 1727. https://doi.org/10.3390/electronics12071727

Graham, C. R. (2009). Blended learning models. In M. Khosrow-Pour (Ed.), *Encyclopedia of information science and technology* (2nd ed., pp. 375–382). IGI Global.

Hall, H., & Davison, B. (2007). Social software as support in hybrid learning environments: The value of the blog as a tool for reflective learning and peer support. *Library & Information Science Research, 29*(2), 163–187. https://doi.org/10.1016/j.lisr.2007.04.007

Hasse, C. (2017). Technological literacy for teachers. *Oxford Review of Education, 43*(3), 365–378. https://doi.org/10.1080/03054985.2017.1305057

Heonjun, Y. (2021). Progress to the metaverse era: 04. New services and cooperation. *Journal of the Korean Society of Mechanical Engineers, 61*(8).

Hernandez-Pozas, O., & Carreon-Flores, H. (2019). Teaching international business using virtual reality. *Journal of Teaching in International Business, 30*(2), 196–212. https://doi.org/10.1080/08975930.2019.1663779

Hetherington, E. M., Parke, R. D., & Locke, V. O. (1999). *Child psychology: A contemporary viewpoint* (5th ed.). McGraw-Hill.

Horovitz, B. (2012, May 4). After Gen X, Millennials, What Should Next Generation Be? *USA Today*. Retrieved on September 30, 2023 from https://www.dentistryiq.com/dental-hygiene/clinical-hygiene/article/16360742/gen-z-a-new-beginning-or-beginning-of-the-end

Horovitz, T., & Mayer, R. E. (2021). Learning with human and virtual instructors who display happy or bored emotions in video lectures. *Computers in Human Behavior, 119*. https://doi.org/10.1016/j.chb.2021.106724

Hrastinski, S. (2019). What do we mean by blended learning? *TechTrends, 63*(4), 564–569. https://doi.org/10.1007/s11528-019-00375-5

Hyun, J. J. (2021). A study on education utilizing Metaverse for effective communication in a convergence subject. *International Journal of Internet, Broadcasting and Communication, 13*(4), 129–134. https://doi.org/10.7236/IJIBC.2021.13.4.129

Jenkins, H. (2009). *Confronting the challenges of participatory culture: Media education for the 21st century*. MIT Press.

Jensen, L., & Konradsen, F. (2018). A review of the use of virtual reality head-mounted displays in education and training. *Education and Information Technologies, 23*(4), 1515–1529. https://doi.org/10.1007/s10639-017-9676-0

Jones, R. H., & Hafner, C. A. (2021). *Understanding digital literacies: A practical introduction* (2nd ed.). Routledge.

Khandelwal, K., & Upadhyay, A. K. (2021). Virtual reality interventions in developing and managing human resources. *Human Resource Development International, 24*(2), 219–233. https://doi.org/10.1080/13678868.2019.1569920

Korucu, A. T., & Alkan, A. (2011). Differences between M-learning (Mobile learning) and E-learning, basic terminology and usage of M-learning in education. *Procedia—Social and Behavioral Sciences, 15*, 1925–1930.

Ladd, G. W., & Mize, J. (1983). A cognitive–social learning model of social-skill training. *Psychological Review, 90*(2), 127–157. https://doi.org/10.1037/0033-295X.90.2.127

Leung, T., Zulkernine, F., & Isah, H. (2018). *The use of virtual reality in enhancing interdisciplinary research and education*. Retrieved on November 15, 2022 from https://arxiv.org/ftp/arxiv/papers/1809/1809.08585.pdf

Lin, L., Ginns, P., Wang, T., & Zhang, P. (2020). Using a pedagogical agent to deliver conversational style instruction: What benefits can you obtain? *Computers & Education, 143*, 10–21. https://doi.org/10.1016/j.compedu.2019.103658

Loveys, K., Sebaratnam, G., Sagar, M., & Broadbent, E. (2020). The effect of design features on relationship quality with embodied conversational agents: A systematic review. *International Journal of Social Robotics, 12*(6), 1293–1312. https://doi.org/10.1007/s12369-020-00680-7

Lucas, G. M., Rizzo, A., Gratch, J., Scherer, S., Stratou, G., Boberg, J., & Morency, L.-P. (2017). Reporting mental health symptoms: Breaking down barriers to care with virtual human interviewers. *Frontiers in Robotics and AI, 4*, 51. https://doi.org/10.3389/frobt.2017.00051

Makransky, G., & Mayer, R. E. (2022). Benefits of taking a virtual field trip in immersive virtual reality: Evidence for the immersion principle in multimedia learning. *Educational Psychology Review, 34*, 1771–1798. https://doi.org/10.1007/s10648-022-09675-4

Makransky, G., Wismer, P., & Mayer, R. E. (2019). A gender matching effect in learning with pedagogical agents in an immersive virtual reality science simulation. *Journal of Computer Assisted Learning, 35*(3), 349–358. https://doi.org/10.1111/jcal.12335

Masuoka, Y., Morikawa, H., Kawai, T., & Nakagohri, T. (2019). Use of smartphone-based head-mounted display devices to view a three-dimensional dissection model in a virtual reality environment: Pilot questionnaire study. *JMIR Medical Education, 5*(1), e11921. https://doi.org/10.2196/11921

Merchant, Z., Goetz, E. T., Cifuentes, L., Keeney-Kennicutt, W., & Davis, T. J. (2014). Effectiveness of virtual reality-based instruction on students' learning outcomes in K-12 and higher education: A meta-analysis. *Computers & Education, 70*(1), 29–40. https://doi.org/10.1016/j.compedu.2013.07.033

Mestre, J. (2002). *Transfer of learning: Issues and research agenda*. Alexandria, VA.

Mitra, S. (2023). Metaverse: A potential virtual-physical ecosystem for innovative blended education and training. *Journal of Metaverse, 3*(1), 66–72. https://doi.org/10.57019/jmv.1168056

Niu, Z., Willoughby, J., & Zhou, R. (2021). Associations of health literacy, social media use, and self-efficacy with health information–seeking intentions among social media users in China: Cross-sectional survey. *Journal of Medical Internet Research, 23*(2). https://doi.org/10.2196/19134

Njenga, J. K., & Fourie, L. C. H. (2010). The myths about e-learning in higher education. *British Journal of Educational Technology, 41*(2), 199–212. https://doi.org/10.1111/j.1467-8535.2008.00910.x

Oh, H. J., Kim, J., Chang, J., Park, N., & Lee, S. (2023). Social benefits of living in the metaverse: The relationships among social presence, supportive interaction, social self-efficacy, and feelings of loneliness. *Computers in Human Behavior, 139*(6). https://doi.org/10.1016/j.chb.2022.107498

Pan, X. (2020). Technology acceptance, technological self-efficacy, and attitude toward technology-based self-directed learning: Learning motivation as a mediator. *Frontiers in Psychology, 11*. https://doi.org/10.3389/fpsyg.2020.564294

Park, S., & Kim, S. (2022). Identifying world types to deliver Gameful experiences for sustainable learning in the Metaverse. *Sustainability, 14*(3), 10–21. https://doi.org/10.3390/su14031361

Peckham, E. (2020, February 25). A multiverse, not the metaverse. Why gaming's virtual worlds are the next stage of social media. *Techcrunch*. Retrieved on March 2, 2023 from https://techcrunch.com/2020/02/25/virtual-worlds-intro/

Peterson, M. (2006). Learner interaction management in an avatar and chat-based virtual world. *Computer Assisted Language Learning, 19*(1), 79–103. https://doi.org/10.1080/09588220600804087

Riches, S., Iannelli, H., Reynolds, L., Fisher, H. L., Cross, S., & Attoe, C. (2022). Virtual reality-based training for mental health staff: A novel approach to increase empathy, compassion, and subjective understanding of service user experience. *Advances in Simulation, 7*(1), 19. https://doi.org/10.1186/s41077-022-00217-0

Salloum, S. A., Al-Emran, M., Habes, M., Alghizzawi, M., Ghani, M. A., & Shaalan, K. (2021). What impacts the acceptance of e-learning through social media? An empirical study. In M. Al-Emran & K. Shaalan (Eds.), *Recent advances in technology acceptance models and theories* (pp. 419–431). Springer.

Saundarajan, K., Osman, S., Kumar, J. A., Daud, M. F., Abu, M. S., & Pairan, M. R. (2020). Learning algebra using augmented reality: A preliminary investigation on the application of photomath for lower secondary education. *International Journal of Emerging Technologies in Learning, 15*(16), 123–133. https://doi.org/10.3991/ijet.v15i16.10540

Schroeder, N. L., Adesope, O. O., & Gilbert, R. B. (2013). How effective are pedagogical agents for learning? A meta-analytic review. *Journal of Educational Computing Research, 49*(1), 1–39. https://doi.org/10.2190/EC.49.1.a

Schwartz, D., Bransford, J., & Sears, D. (2005). *Transfer of learning from a modern multidisciplinary perspective*. Information Age Publishing.

Shamsudin, N., Mahmood, N. H. N., Rahim, A. R. A., Mohamad, S. F., & Masrom, M. (2018). Virtual reality training approach for occupational safety and health: A pilot study. *Advanced Science Letters, 24*(4), 2447–2450. https://doi.org/10.1166/asl.2018.10977

Shao, C., & Hazel Kwon, K. (2021). Clicks intended: An integrated model for nuanced social feedback system uses on Facebook. *Telematics and Informatics, 39*, 11–24. https://doi.org/10.1016/j.tele.2018.12.003

Siyaev, A., & Jo, G.-S. (2021). Towards aircraft maintenance metaverse using speech interactions with virtual objects in mixed reality. *Sensors, 21*(6), 10–21. https://doi.org/10.3390/s21062066

Suh, A., & Prophet, J. (2018). The state of immersive technology research: A literature analysis. *Computers in Human Behavior, 86*, 77–90. https://doi.org/10.1016/j.chb.2018.04.019

Sylaiou, S., Kasapakis, V., Gavalas, D., & Dzardanova, E. (2020). Avatars as storytellers: Affective narratives in virtual museums. *Personal and Ubiquitous Computing, 24*(6), 829–841. https://doi.org/10.1007/s00779-019-01358-2

Tan, S. (2023). Harnessing immersive technologies for innovation in teaching and learnings. In K. Rajaram (Ed.), *Learning intelligence: Innovative and digital transformative learning strategies* (pp. 305–334). Springer.

Torres, I. M., & Briggs, E. (2007). Identification effects on advertising response: The moderating role of involvement. *Journal of Advertising, 36*(3), 97–108. https://doi.org/10.2753/JOA0091-3367360307

Valkenburg, P. M., & Peter, J. (2008). Adolescents' identity experiments on the internet: Consequences for social competence and self-concept unity. *Communication Research, 35*(2), 208–231. https://doi.org/10.1177/0093650207313164

Vygotsky, L. (1978). *Mind in society: The development of higher psychological processes*. Harvard University Press.

Wang, D., Xu, L., & Chan, H. C. (2015). Understanding the continuance use of social network sites: A computer self-efficacy perspective. *Behaviour & Information Technology, 34*(2), 204–216. https://doi.org/10.1080/0144929X.2014.952778

Wang, M., Yu, H., Bell, Z., & Chu, X. (2022). Constructing an edu-metaverse ecosystem: A new and innovative framework. *IEEE Transactions on Learning Technologies, 15*(6), 685–696. https://doi.org/10.1109/TLT.2022.3210828

Webster, R. (2016). Conceptual knowledge acquisition in immersive virtual learning environments. *Interactive Learning Environments, 24*(6), 1319–1333. https://doi.org/10.1080/10494820.2014.994533

Wu, B., Yu, X., & Gu, X. (2020). Effectiveness of immersive virtual reality using head-mounted displays on learning performance: A meta-analysis. *British Journal of Educational Technology, 51*(6), 1991–2005. https://doi.org/10.1111/bjet.13023

Yang, E., & Ryu, J. (2021). Do we perceive emotional gender gesture of virtual avatar as intended? *International Journal for Educational Media and Technology, 15*(1), 10–21.

Zhai, X., Chu, X., Wang, M., Zhang, Z., & Dong, Y. (2022). Education metaverse: Innovations and challenges of the new generation of internet education formats. *Meta, 3*(1), 1–13. https://doi.org/10.54517/met.v3i1.1804

Zhang, K., Suo, J., Chen, J., Liu, X., & Gao, L. (2017). Design and implementation of fire safety education system on campus based on virtual reality technology. *Federated Conference on Computer Science and Information Systems*, 1297–1300. https://doi.org/10.15439/2017F376

Zhou, B. (2022). Building a smart education ecosystem from a metaverse perspective. *Mobile Information Systems, 3*, 1–10. https://doi.org/10.1155/2022/1938329

Zimmerman, B. J. (1995). Self-efficacy and educational development. In A. Bandura (Ed.), *Self-efficacy in changing societies* (pp. 202–231). Cambridge University Press.

Chapter 8
Deception in the Metaverse

Deception is a perpetual risk in the metaverse. It encompasses fake news, misinformation, and disinformation (Petratos, 2021). Deception in the metaverse should be understood as misleading content or tactics aimed at generating public or personal harm or making a profit from the victims (e.g., through identity theft). Within the context of this chapter, deception comes from either state or non-state actors who disseminate content or employ strategies without too much effort or too many obstacles. A major problem is that this type of communication is difficult to control because it is often decentralized. It cannot be blocked with traditional centralized tools. Under these circumstances, a lack of awareness or inadequate supervision will open doors to various security attacks (Fraga-Lamas & Fernández-Caramés, 2020).

We know by now that user-to-user communication is the foundation of the metaverse. This virtual environment is essentially a platform where any participant can interact with any other. The problem is compounded by the fact that such participants can communicate with others via avatars and not with their real selves, causing societal rules for interactions to vanish. Alleviation of activities and interactions in the metaverse is one recommendation, but can be problematic if it becomes too massive, even if the objective is to keep the metaverse safer from deception, fake news, misinformation, and disinformation. Otherwise, a large proportion of users may show less interest in using the metaverse, thereby leading to losses for big-tech companies. Moderation at such level is also easier said than done. Currently, moderation of any virtual environment poses a challenge and cannot curtail the aforementioned problems (Huq et al., 2022).

In line with these contentions, it is anticipated that deception in the metaverse will continue to grow because of the massively available immersive technologies that can be easily tweaked to abuse interactants, "fool" their senses, and keep them dormant to the moral ramifications of deceptive methods. What is hazardous to metaverse users is that the platform can be maneuvered and controlled by big-tech companies in ways that are deceptive in and of themselves. This is already the case

J. Matusitz, J. Dacas, *Communication in the Metaverse*, Signals and Communication Technology, https://doi.org/10.1007/978-3-031-63485-7_8

with social media platforms, where aggressive tactics and engagement strategies are being used to profit from users (Bibri, 2022). Aside from big-tech companies, propagandists and extremists always explore new methods to engineer fake content while eschewing detection. They are malevolent actors deliberately working toward taking advantage of emerging systems' weakest elements, thereby making the metaverse an ideal "place" for deception. After all, the more people can be reached, the more people can be not only deceived but also radicalized or recruited (Bajwa, 2022). In this communication dynamic, teeming with anonymous senders and receivers, an astronomical volume of content disseminated in the metaverse as "information" is questionable and, in too many cases, intentionally misleading—i.e., the example of fake news mentioned earlier (Zhang & Ghorbani, 2020). Ultimately, malevolent actors can quickly diffuse malicious content to shape people's cognitions, attitudes, and behaviors (Obadă, 2019).

Deception: Definitions

Deception is the act of intentionally transmitting content to a receiver in order to mislead that receiver (Ekman, 1992). It happens when communicators manipulate the information contained within their messages to share or impart meanings that diverge from the truth (Buller & Burgoon, 1996). The diffusion of digital misinformation has been labeled a considerable threat to democracy. Communication, psychology, social science, and computer science scholars are examining the deep-rooted causes for the viral propagation of misinformation, in spite of the fact that the metaverse continues to deploy countermeasures (Shao et al., 2018).

The Ability to Detect Falsity

A communicator may have a range of motivations to produce a sense of falsity that can trigger either positive or negative reactions from the audience, leading to additional corollaries. Senders may deceive receivers by diffusing false information with the intent to change or embolden receivers' emotional, behavioral, and cognitive responses. Falsity comes in many shapes or forms. It can negatively impact the receiver or the actual topic that is targeted by the falsity (Park et al., 2020). When subjected to falsity, receivers can react by expressing concerns regarding the verisimilitude of the message or interaction. Verisimilitude is the extent to which something is regarded as real (Campbell et al., 2022). Verisimilitude can manifest itself through truthiness (the extent to which something is true based on how it looks like) (Berthon & Pitt, 2018) and authenticity (the extent to which something is factual, related to the subject matter, accurate, legitimate, original, and adroitly communicated) (Nunes et al., 2021). Perceptions of verisimilitude enhance the receiver's capacity to identify falsity when it is apparent or perceived to occur during the interaction. The ability of a receiver to evaluate the degree of truthiness is mostly based on how much he or she trusts the sender and on confirmation bias—a phenomenon whereby a person interprets, privileges, and recalls information that confirms preexisting beliefs (Berthon & Pitt, 2018). This can lead to potentially false interpretations too (Plangger & Campbell, 2022).

Fake News

Fake news refers to (intentional) disinformation or (unintentional) misinformation disseminated in traditional venues or on virtual platforms. In many cases, it is intentional because it is written or posted with the purpose of misleading audiences. Eventually, the message originator seeks to obtain desired results—financially, politically, ideologically (i.e., propaganda), or even personally (e.g., by grabbing people's attention). The term "fake news" is not new. It seems to have been popularized in the late nineteenth century, although analogous terms such as "false news" have been employed since the sixteenth century. This is why the potential for people to distort information for a variety of motives has long been understood (Mason et al., 2018). As we form an impression of content in the metaverse, we become exposed to a bulk of false or misleading news, ruses, conspiracy theories, click-bait talking points, junk science, and even jokes (Vosoughi et al., 2018).

Such content is collectively described as "disinformation." The financial incentives through advertising are clear and ubiquitous, but political, ideological, and personal motives can be equally consequential (Mustafaraj & Metaxas, 2010). While the veracity (or lack thereof) of disinformation is not always easy to prove, its destructive potential has been demonstrated in not only communication and psychology but also health and finance. Part of the reason why we should worry about its destructive potential lies in its effect on believability (Hotez, 2016). Believability is the extent or scale of believing, which measures the users' probability to believe and act on fake news after encountering it (Kumar et al., 2021). Kim and Dennis (2019) contend that "believability … [of news] can affect the action of users." Put another way, user behavior is heavily shaped by the believability of content in the metaverse.

Misinformation Vs. Disinformation

Misinformation is involuntary or unpremeditated false or inaccurate information. Thus, it is spread without ill intentions but, nevertheless, clashes with the epistemic consensus of the scientific community, or any type of well-meaning community at a given time (Scheufele & Krause, 2019). Misinformation can be partially or totally incorrect. Even information that is not created for any political or ideological purpose can still be detrimental. An illustration of this was evident in the high volume of online fake news that contributed to the Brexit general vote by the electorate in the United Kingdom in 2017, called the "Leave" campaign. The campaign utilized a spectrum of inaccurate talking points—from warnings about the UK's invasion by refugees and asylum seekers to unfounded information that a whopping £350 million a week was given to Brussels (the headquarters of the European Union) by the UK Parliament (money that would allegedly be saved if the Brexit referendum prevailed) (Smith, 2017). During the same year, the World Economic Forum published a warning on the potential of cyber news to alter the perception of reality. This anecdote is the type that not only poses a risk to democracy, but it is also spawning a paradigm shift in the way people consume information (Cinelli et al., 2020).

On the other hand, disinformation is intentionally inaccurate information with the purpose of manipulating or harming the audience (Allcott & Gentzkow, 2017). The reasons for disinformation are plentiful, including financial motives, political power, creating conflict in society, or even stirring international tensions. It is frequently based on sociopsychological reasons—the aspirations to create social or political change, as would be the case when a passionate individual seeks to spread new beliefs or values (with respect to social issues, religion, or more detailed issues such as climate change and vaccination). Disinformation is a category of mal-information grounded in information twist or under the guise of reinterpretation of content (Repez & Popescu, 2020). It can also manifest itself through hostile social manipulation, referred to as "the purposeful, systematic generation and dissemination of information to produce harmful social, political, and economic outcomes in a target country by affecting beliefs, attitudes, and behaviors" (Mazarr et al., 2019, p. 6). Hostile social manipulation focuses on beliefs and attitudes, as opposed to physical assets or military forces. To be more precise, the target is the opponent's will (Mazarr et al., 2019).

Conspiracy Theory

Both misinformation and disinformation can be the byproduct of conspiracy theories. A conspiracy theory is a belief that a phenomenon has been secretly engendered by powerful people (Walker, 2014). It is a subset of a belief system that commonly embroils a malevolent force (e.g., a government body or secret elitist group) that seeks to mastermind events; this can be done by providing disinformation concerning the details of such events to the public, partly as a plot toward accomplishing an ominous goal (Brotherton et al., 2013). Conspiracy theories advance some ideologies that become false consciousness because they are taken as true by a certain percentage of the population. Conspiracy theories proffer agendas not accessible to unwitting audiences or through "surface" discourses (White, 2002).

A conspiracy theory is to be differentiated from a conspiracy. Unlike the latter, the former denotes a hypothesized conspiracy with peculiar characteristics, including but not limited to resistance to the mainstream powers-that-be or even society at large. It also consists of forming an opposition to established scientific consensuses (i.e., the validity of vaccines) or historical consensuses (Brotherton, 2013). Some studies (e.g., Barron et al., 2014; Freeman & Bentall, 2017) put forward the notion that conspiracist ideation—the belief in conspiracy theories—is psychologically detrimental or pathological. Conspiracist ideation has been shown to be correlated with a lower degree of analytical thinking, low intelligence, psychological projection, a victim complex, a delusion complex, and Machiavellianism.

Selective Exposure Theory

In the metaverse, participants may operate in filter bubbles that do not receive opposing viewpoints and content challenging their social groups' positions. This is a throwback to the concept of confirmation bias; individuals affected by it tend to favor information with which they identify or concur. Magnified by digital algorithms, this occurrence polarizes social groups of participants in the metaverse even more. This situation can also be explained through the notion of motivated reasoning, a psychological dynamic according to which people have a tendency to believe content only when they want it to be true. In the realm of political campaigns, if a person's ideal candidate lost an election, which causes the disappointed voter to spread the false claim that the candidate "actually won," other users could jump on this bandwagon of disinformation and cause the phenomenon of motivated reasoning to grow even more (Epley & Gilovich, 2016). This daily occurrence can also be described by the main tenets of selective exposure theory.

Widely applied in communication studies, selective exposure theory rests on the premise people who spend time on either traditional or digital media are likely to believe content that reinforces their views, positions, attitudes, and values, at the same time rejecting any conflicting information (McLennan, 1970). According to this theory, such individuals tend to expose themselves to content that they see as favorable, while ignoring unfavorable items that circulate on platforms. Selective exposure can take several forms: (1) selective exposure to content and (2) selective exposure to certain media of communication—as it is sometimes the case that even the most educated ones will avoid using social media or reading internet content because of negative past encounters with fake news or conspiracy theories (see Jia & Johnson, 2021; Melki et al., 2023).

Stroud (2018) provides a few explanations as to why selective exposure occurs frequently among people, even in present-day society (an epoch during which people are more educated and have much wider access to information than yesteryears). One explanation is that selective exposure emerges when someone is exposed to contradictory ideas or information, and to resolve this issue, will favor information rooted in one's preexistent beliefs. Another explanation is that absorbing familiar or agreeable content decreases the amount of cognition necessary to process authentic and factual content. This means that people will devote less time and energy when believing like-minded information because it is, cognitively speaking, easier (Stroud, 2018).

The Dangers of False Information in the Metaverse

Apprehension about the widespread presence of false information has not abated because the properties of the metaverse smooth the way for the relatively inexpensive production of deceptive information and the ease of diffusing it across platforms (Weeks & de Zigak, 2019). Both human users and avatars are known to contribute misinformation or disinformation in user-generated content (UGC) mode, as illustrated with the sharing of unreal experiences in virtual environments, for all sorts of reasons (Guo et al., 2017). This can also happen when false information is sent through uncalibrated sensory devices, which tend to generate low-quality data, ultimately undermining users' experiences in the metaverse (Ali et al., 2022).

Deception in the metaverse can be far-reaching because the range of technologies that allow it place information manipulators in a digitally advantaged environment where anyone not only sees but also hears and feels such deception. In VR environments, people can easily relate emotionally with others, even when they find themselves navigating through profound disinformation. The science examining these psychological phenomena continues to evolve but, so far, results reveal that they are powerful (Han et al., 2022). Another consideration of deception in the metaverse is the difficulty in identifying propaganda, especially when manipulated content does not always happen through words (but also through images and sound). Although users might be able to recognize signs of manipulation in some instances—e.g., noticing bizarre blinking patterns or strange-looking lip movements in a deepfake or detecting unsuitable modulation of an automated voice in a video—they could have a more difficult time identifying such tells when different senses are cleverly merged with movements in other immersive media situations (Woolley, 2020).

The Danger of Anonymity

The anonymity that the metaverse affords can give rise to concerns about how information is transmitted digitally. We have already been exposed to many people's concerns about fake news and how bad virtual communications can be when sharing data and other categories of knowledge. In the three-dimensional world, people traditionally gain knowledge by means of rule-based analyses. False information has negative reverberations on the culpable and/or the gullible who share it with others, in spite of good intentions. In the metaverse, trust and reliability should have as much value and play a critical role. However, in many situations, these values are hard to ascertain because the interactant may not be a real person, but an anonymous avatar. On a side note, it might be worthwhile to explore other reaction principles, such as punishment, to understand why absence of trust is as bad as false belief. Incentive designs to raise the level of trust among anonymous players should be essential to functionality (in order to reduce the occurrence of deception) (Fernandez

& Hui, 2022). Hence, anonymity clears the way for the diffusion of content, whether it is accurate or fake. With only a small latitude to distinguish between the two, it is up to the participants in the metaverse to find the person or content reliable or not.

False information impedes the free flow of public opinion because of the manifold methods of abusing anonymity and, thereby, manipulating users. To this point, a major impact is one that widens the chasm between deceptive power and active citizenry. Additionally, posting content has never been easier, thanks to all sorts of UGC software programs. Thus, the ability of AR and VR to imitate reality is both promising and harmful, because many immersive experiences do not fall in a category of rational thinking and, as a result, aggravate the effects of likely manipulations (Fraga-Lamas & Fernández-Caramés, 2020).

The Danger of Virtual Humans

In the metaverse, virtual humans can be regulated on the computer screen or through devices such as head-mounted displays; this is why they should be identified as embodied agents rather than avatars. The notion of the virtual human is rather simple to understand once the user knows how to navigate its functions through a computer or digital device, independent of the quality of computer graphics, prerecorded animations, or speech production (Brown et al., 2023). In fact, even with virtual human models created at the turn of the century, behavioral algorithms for nonverbal animation were already so advanced that a certain number of random people would fail the Turing Test today. That is, they would not be able to tell whether such older virtual human was controlled by a computer or a real person (at least, in short interactions) (Bailenson et al., 2004).

In a deceptive environment, metaverse creators can get "people" to inhabit/occupy scenes and create the illusion of consensus. Yet, unlike bots on social media, these embodied agents can be nonverbally excellent—as they would be verbally—when interacting with users. Early work by Blascovich (2002) revealed that embodied agents frequently give the impression of conformity. In a virtual gambling scene, people altered their betting attitudes when embodied agents were cleverly placed in the scene; more precisely, when agents bet high or low, the users followed suit with similar bet sizes. In other words, embodied agents are more persuasive than others types of online personas, such as simple agents (Kantharaju et al., 2018). In the metaverse, populating a scene with a myriad embodied agents is easy, so deception reigns supreme, and will not likely disappear. Moreover, such virtual humans can be implanted within a scene to enhance demographic similarity to users, as studies have demonstrated that users are more likely to endorse a policy initiative when embodied agents of the same gender discussed the policy, in comparison with virtual humans of the opposite gender (Guadagno et al., 2007).

The Danger of the Editability Affordance

A major affordance of the present-day digital world is editability (Weber & Haseki, 2021). Editability refers to the degree to which users perceive the digital world as an "opportunity producer," one that enables them to manipulate content that is posted, reposted, or shared, on a platform (Pee, 2018). This affordance allows any user to edit or revise his or her own content. Enter Facebook. At the lowest level, "Facebook allows users to edit descriptions of their posts or even delete contents published on their walls" (Chan et al., 2019, p. 586). This is an affordance that gives Facebook users the ability to take back their acts of posting and/or sharing false information by maneuvering things (e.g., erasing, editing). In another example, to avoid discomfort, VR users can edit or remove their posts or photos when the content is sensitive or pertaining to controversial issues (e.g., COVID-19)—perhaps because they found the content to be false or wrong. For platform custodians, this affordance makes tracing false content difficult; consequently, deception is easy (Wong, 2021).

The fact that people today rely predominantly on edited content that leads to errors, manipulation, and mass movements of moral panic is undeniable. It is an information storm that operates as an irrational narrative; the problem is not only about deception in the metaverse but also about emerging divisions among users. False content thrives on people's inability to process the colossal information wave that they have to manage on a day-to-day basis. This incapacity causes us to let ourselves influenced by other users' editing or tweaking of data flow. This is often the case for those segments of content that most people do not examine more closely or for whose explanation we do not possess adequate critical thinking skills. Therefore, it is more difficult to validate such content as true or false (Repez & Popescu, 2020).

The Danger of the Association Affordance

In the metaverse, association affordance denotes the wide availability of connections that can be formed between individual users and their content (Treem & Leonardi, 2013). It is the degree to which a user believes that the virtual environment "offers the opportunity to share responsibility for his or her post with other users who interact with the post on the platform" (Chan et al., 2019, p. 586). This latter description also alludes to the denial of association and associated accountability or, to better put it, association-denial affordance. This affordance enables users (1) to deny any accountability for being associated with deceptive content and (2) to deny any responsibility for diffusing that content. Instead, responsibility is passed on to others who interacted with that user (through shares, likes, or comments). In turn, these users can also deny any responsibility and transfer it to others; so down the line… Naturally, users can simply claim that they did not create any content (but were merely sharing it). The association-denial affordance not only complicates any type of user responsibility, but it also makes deception easy, thereby allowing false content to flourish even more. The easiness of such phenomenon can be attributed, in part, to the method by which content is processed or selected from the high volume of data flows (Hossain et al., 2022).

The Theory of Weak Ties

Another important consideration is the rapidity of diffusion of false content as a result of tie strength. Tie strength is "the potency of the bond between members of a network" (Mittal et al., 2008, p. 196). These bonds can be categorized as strong ties or weak ties depending on the significance associated with the relationship (Ibarra et al. 2005), the frequency of social contacts, and the degree of intimacy and reciprocity between two people. Therefore, strong ties occur when there are higher degrees of connectedness, reciprocity, and emotional attachment than weak ties (Granovetter, 1973). Content can be disseminated more easily in small and tight networks of mutual relationships (Kadushin, 2012), such as those tightly-knit connections between close relatives and high-school friends on social networking sites. Circumstances are different in wider and less tight networks, such as those situations in which people develop a one-directional rapport with famous personalities, politicians, and journalists on platforms such as Twitter (González-Bailón & Wang, 2016). This type of remote audience is a theoretical audience that represents the entire user base of the platform, rendering user connections weak, one-directional, and without real in-person contacts (Theocharis et al., 2021).

An important aspect of the metaverse deals with communication, interaction, and immediacy; this is where the broad spectrum of information and data comes into play. Authenticity means not diffusing false content, not producing false narratives. In metaverse communities, weak ties can be as bad as strong ties for two reasons. First, users can trust content that even comes from people that they do not know. Second, it could be a two-step process in the sense that false content can, at first, come from sources with whom we have no genuine connections. Yet, such false content, in turn, gets transmitted across networks, including those where strongly interacting individuals have formed tightly-knit groups. People have a strong tendency to trust information that comes from those around them (Larson, 2017).

According to Lazarsfeld, Berelson, and Gaudet (1968), traditional mass communication generally does not directly influence the audience's opinions because it is merely what they see on television or what they read in the newspapers. These three authors suggested two stages of influence. The first stage was the one from traditional mass communication. In the second stage, people tend to believe the opinions expressed by those around them. However, in the metaverse, even strangers with weak links to one other can shape their beliefs—i.e., the strong power of influencers. Whereas strong ties of kinship and friendship induce sentiments of belonging and safety, weak ties with strangers (or even mere "acquaintances") can unleash the infinite universe of ideas and opportunities. This phenomenon can be a force multiplier through further strong and weak ties in the digital world, make global campaigns of deception, misinformation, and disinformation spread such as wildfire (Szűts, 2021). The universalization of content comes with a heavy price because it constantly opens Pandora's boxes of new challenges.

Moral Panic Theory

One of the profound consequences of the volume of false content in the metaverse can be explained through moral panic theory. Moral panic denotes a mass movement, based on true, false, or exaggerated information, about an individual, a group, or a nonhuman subject (Rohloff & Wright, 2010). It is formed on a widespread alarm or angst, usually irrational, because of a perceived threat (physical or symbolic) to society (or society's mores, as determined by the communication media that were used) (Cohen, 1972). Put simply, moral panic is an exaggerated response of public concern, caused by a communication medium, over the values and/or safety of a society. In addition to public concern, it leads to mass hysteria because the public believes whatever content was reported on what transpired vis-à-vis that individual, group, or nonhuman subject (Rohloff & Wright, 2010).

As Rohloff and Wright (2010) continue, "'moral panic' is a sociological concept that seeks to explain a particular type of overreaction to a perceived social problem" (p. 404). According to Cohen (1972), the mastermind of the moral panic framework, the concept of "moral panic" constitutes a situation in which media has fashioned a "folk devil," and the public demands that the powers-that-be take decisive actions about it. The metaverse is a key illustration of this theory, because players can turn the virtual environment in a way that causes observers to think it functions as an instigator of division, antagonism, malevolence, and panic (Walsh, 2020). In many cases, the digital world can present information in a manner that challenges mores and social norms, causing people to enter into a panic mode, which is often irrational (Hossain et al., 2022).

Moral panic theory can be explained in more details through five sequential stages (Crossman, 2019). First, someone or something is portrayed or framed as a threat to the values or safety of society. It is a situation in which the public overreacts to the actual threat that "someone" or "something" poses. Second, this threat is presented in the media in a fairly recognizable and simplistic fashion. Studies report that the notion of spreading false content, myths, and rumors in the virtual world skyrocketed during the COVID-19 pandemic (Al-Zaman, 2021). Third, a rapid and extensive public concern forms as a result of the method that the media uses to portray the threat. During the pandemic, mandatory lockdowns and its associated guidelines initiated a whole range of deceptive content on digital platforms across the world, fueled by mass anxiety (Al-Zaman, 2021), fear, and panic. This caused panic buying of essential hygienic and food items (Ahmad & Murad, 2020; Arafat et al., 2021). Fourth, as a reaction to such chaos, the appropriate states and policymakers institute new laws or guidelines. To keep panic behaviors under check, governmental agencies adopt strict measures, ranging from legislative action and heightened public awareness to the struggle with false content (Robinson et al., 2019). Lastly, in due time, the panic withers away or brings about social changes—whether or not these social changes are made for the benefit of society (Hossain et al., 2022).

Social Engineering in the Metaverse

Social engineering is a concept used for a wide range of malevolent activities conducted through deceptive human interactions. It relies on psychological manipulation to mislead users into believing specific content, committing errors, or disclosing sensitive information. To this last point, the purpose is to maneuver people into giving away valuable and sensitive data to satisfy the deceivers' or criminals' needs (Salahdine & Kaabouch, 2019). Social engineering occurs in two major steps. First, the perpetrator looks up details on the targeted victim to amass necessary background information, such as useful points of entry and fragile security protocols, required to proceed with the subterfuge operation. Second, the perpetrator moves to acquire the victim's trust and provide incentives for subsequent actions that lower the victim's guard; an example would be divulging sensitive information or giving access to critical resources (Kalniņš et al., 2017).

The information age has pervaded our lives, the metaverse being the most important communication and information exchange medium. On a day-to-day basis, communication gets distributed over a wide range of online communication channels. At levels lower than Web 3.0, Web 2.0 services available on social networking sites have become second nature to most users; they are a part of our everyday routine in private and professional communication. People are more likely to trust other people in comparison with trusting computers or machines. Consequently, people form the weakest link in the security chain. Ill-natured activities effectuated through human interactions have an impact on a person psychologically to release confidential information (Hadnagy, 2018).

How Does It Work?

Social engineering attacks are distinguished through the use of several techniques employed against victims: (1) reciprocity (a commodity or a service is offered and given, asking then for another one in return), (2) authority (the manipulator presents him- or herself as an important person, such as a law enforcement agent), (3) urgency (a hypothetical situation is produced to urge the victim to take action imminently, such as the situation that happened with the WannaCry cryptoworm in 2017, which afflicted 230,000 computers in 150 nations), (4) sympathy (oozing a sense of confidence and optimism), (5) concession (admitting to flaws so as to appear more human, as would be the case when the perpetrator has to deliver a card that should have been renewed a whole ago, at the same time directing the victim to a fake website), and (6) preloading (a practice that generally consists of sending general news about some situation and then creating an opportunity about it, directing the victim to a link that has just been sent) (Pajuste, 2022). In contrast to terrorist attacks conducted online (see Matusitz et al., 2019), social engineering does not entail physical violence, and is not so abrupt or direct. This is why it is called malevolent deception.

In Relation to False Information

It is clear, by now, that the affordances of the metaverse, including anonymity, constitute a massive motivator for deceivers to create subterfuges and spread false content. In particular, misinformation and disinformation spaces can be transformed into VR honeypots for intelligence gathering. Perpetrators can diffuse fake content by performing meticulous sentiment analysis, such as the communicative tactics employed by politicians to sway public opinion. Sentiment analysis is accomplished by examining personal data gathered in the metaverse—through information gathering, but also eye tracking, body language reading, location and movement tracking, voice tonality, and so forth. More precisely, criminals or state actors attempt to identify groups of individuals who are vulnerable to certain topics and then apply targeted narratives to sway them into taking particular actions (Huq et al., 2022).

With the assistance of social engineering techniques, a set of actions could be easily generated toward the social cancellation of a victim who has not engaged in any offense. An illustration of this would be a case in which AI and information from social networks are used to build a smear campaign through fake news about a specific individual. That specific individual is portrayed as a rival to be defeated by an adversary or competitor. The social engineers propagate false information by initially obtaining vital data from the victim. Such data can be obtained by particular deceptive techniques, such as flattering the victim through empathy; posting uncertain information to be rectified by the victim; demonstrating a fake ignorance about something that the target controls; obtaining useful information through the explanation provided by the victim; acting as a sounding board to the victim by telling him or her personal stories (so that he or she tells the perpetrator theirs); and formulating clever questions (which could be closed- or open-ended questions). Prefabricated scenarios are generally based on a set of protocols and brilliant staging, research, and planning. Of equal relevance for successful social engineering is a solid understanding of body language (i.e., by projecting confidence, avoiding sweats, and exuding proper body movements) and understanding expressions (with proper positioning and imitation) (Pajuste, 2022).

Recommendation Algorithms

A recommendation algorithm is an AI algorithm that uses a great deal of machine learning and "big data" to suggest or recommend valuable content to users or consumers. Such AI design will source data in multiple ways to offer suggestions to users. The latter can be reached after entering queries in popular search engines or by inadvertently giving information to malicious parties (Shu et al., 2018). Recommendation algorithms are utilized by the majority of online platforms, including smartphone apps and similar devices. Customized recommendations are the foundation of the new augmented actors in the metaverse. They are efficient because, through various sensors, they can put their fingers on our desires and adapting them to each situation. This would be acceptable if the aims were noble. Naturally, there are two sides to a coin. The negative corollaries of recommendation algorithms can be disastrous because of the exponential growth of social engineering schemes around us (Bojić, 2022).

Naïve trust in recommendation algorithms poses a grave concern. AI-based tools are a keystone of the metaverse and responsible for a large chunk of advertisements to users. These algorithms are increasingly replacing other apps by gaining our attention and directing it to other similar platforms. They are also set up in a way that causes us to lengthen our virtual usage time, which suits the interests of social engineers. After all, the more time we spend in the digital world, the more likely we fall prey to all sorts of attacks. Furthermore, by recommending like-minded content to like-minded people, recommender systems form echo chambers, promoting false content and jeopardizing the democratic process. We have seen how vulnerable humans can be and how easily deceived people can be in believing misinformation or disinformation. AI algorithms operate in the same way as mass media when it comes to pressuring audiences, generating even more dependence on digital or traditional media. As time passes, this process of indoctrination and resultant addiction becomes more and more efficient (Bojić, 2022).

Injection of False Data and Data Tampering

It is not unusual for social engineers to inject false details; for example, by giving the wrong instructions and false data in order to mislead diverse types of entities (Liang et al., 2017). AI-created content makes it easy for users to be immersed during the early stages of the metaverse. Attackers are aware of this and may infuse poisoned gradients or training samples during an AI training session. One objective of such social engineering is to generate biased AI models. The false instructions (or feedback) endanger the safety of users or their physical equipment. In a specific example, by showing the level of pain expressed by the users in any metaverse game (e.g., pain for being shot), the user might unintentionally expose him- nor herself to severe consequences or, in worst-case scenarios, might give rise to even more effects (Ali et al., 2022).

In regards to data tampering, the social engineering process operates through identity features that are designed efficiently monitor and identify changes in data during all sorts of interactions. The attackers may delete, replace, fabricate, or alter raw data over the course of a user's profile usage in the metaverse. In doing so, the attackers tamper with the users' activities (whether through avatars or various apps) (Shu et al., 2018). By extension, the attackers might remain invisible when tampering with users' data or might be able to cover their tracks during their illicit activities in the virtual environment (Ali et al., 2022).

Case Study: The Zero-Trust Architecture as a Solution?

To curtail the problem of deception and disinformation in the metaverse, Gupta et al. (2023) have proposed the Zero-Trust Architecture (ZTA) model. Owing to the high volume of stakeholders devoting their time and energy to deliver opportunities for the metaverse experience, it is primordial that an open metaverse adopt the ZTA model, which requires user/entity verification for all activities (unlike many other current models on digital platforms). This process was set in place to guarantee accountability for user actions. User/entity verification can be done through multifactor authentication, data validation, audit trails and provenance (by means of blockchain), tight access control, privilege supervision, constant surveillance of user/entity behavior, trust, reputation management, and any appropriate disciplinary action against problematic users/entities (Alevizos et al., 2021; He et al., 2022).

Step One: Verification and Registration

One of the main difficulties in dealing with problematic user behavior in the metaverse is the widespread anonymity on virtual platforms. The use of bots is also widespread in the use of deception. Hence, the ZTA model posits that requiring user/entity verification should be mandatory in the metaverse. In other words, only preapproved users/entities should be allowed to register for it; in doing so, it would attenuate the challenge linked to existing platforms. In 2022, Twitter enforced the "blue tick" symbol on validated user profiles as proof that the user's identity had been verified and authentic. This provided a guarantee that true content creators could earn credit for their content and masqueraders would not. The ZTA model rests on the same principle to confirm that all users and entities have had their identity and antecedents approved before embarking on the metaverse to perform their activities (Nguyen et al., 2022).

Step Two: Identity Management and Authentication

At this stage, the level of authentication is done before giving users the permission to access the metaverse. Besides performing authentication, this stage also handles identity management. In the metaverse, users can either have their identities publicly visible or use an avatar or some form of digital twin/persona to access its services. Therefore, the identity management operation is set up to ensure that the creation of an avatar by the user is legitimate; the user also has the option to store the mapping of his or her avatar in an encrypted database (in this manner, user privacy is ensured). Another possibility for delivering trusted identity management is the use of a reliable third-party, which can map the real identity of the permitted user with his or her own digital twin. In this fashion, the platform itself is not apprised of the actual mapping (cited in Gupta et al., 2023).

Step Three: Access Management

The access management layer consists of ensuring coherent or uninterrupted access to the VR-rendering layer and to the metaverse. In essence, this stage encompasses the access policies for all the entities involved in the metaverse so as to provide modulized (i.e., highly detailed) access control at the level of each entity. Associated with the access policies for every entity are the privilege lists, which determine the rising privilege levels based on the acquired reputation by that entity (Zheng et al., 1998). At the individual level, especially when trying to detect a social engineering threat, it is better to pay particular attention to the malevolent actors, who have been trained in cleverly conveying control and trust, obtaining or giving favors and gifts, using humor, and giving seemingly valid reasons for being granted access to information (Pajuste, 2022).

Step Four: Control, Trust, and Reputation Management

The access and privilege policies, guidelines, and regulations are arranged and established into the rules and regulations apparatus, which incorporates numerous policy/rules-enforcer agents. User consent in obtaining particular areas/content, as well as granting other users access to the personal space of users, is encompassed by the consent engine. In the metaverse, the mutually agreed-upon participation and interactions between different entities are constantly surveilled by metaverse guardians, which operate in concert with the policy/rules-enforcer agents to identify any violations or errant actions by any of the entities. All violations are put on the blockchain's "radar," thereby adjusting trust and reputation scores accordingly. "Repeated offenders" may see their access privileges revoked or may get demoted to lower levels. Reputation scores are set for entities over a lengthy period of time. As in the traditional world, higher scores open doors to higher access privileges for users and entities (Jung & Jeon, 2022).

Other Solutions

In addition to the four steps of the ZTA model, other solutions should be made available. To begin, to combat deception, misinformation, and disinformation in the metaverse, detection software and procedures put in place to locate and isolate false content have spawned a relatively new research domain called *Cognition Security* (*CogSec*), which examines the effects of anything that falls under the overarching problem of false information, particularly when it exerts negative influence on human cognition. In this context, we could evoke the concepts of misperception, unknown acquisition of cognitive elements, and adapted and guided formation or change of opinion (with the aim of modifying behavior). These concepts contribute to the formation of a biased attitude in decision-making. So, the purpose of *CogSec* is to enact measures that fight back against these effects. Prior studies on cognition and memory have attempted to explain how this category of information impacts people and the reasons that cause them to be so easily prone to believing the content (Repez & Popescu, 2020).

The rapid increase in the availability of digital tools and the volume of digital content require innovative solutions to manage such challenges which, theoretically, will confer users more skillfulness in detecting false information. At this stage, it is hard to validate the authenticity of information by merely taking a closer look at images, signals, or texts—particularly because of the heterogeneity of these modalities. Thus, it would be more advisable to train users by combining the attributes from these modalities and learning the characteristics peculiar to each of them. Furthermore, on account of the unstructured types of languages found in the metaverse, and the complexity of the human recognition process, it is difficult to concentrate on the essential words (or the corresponding images) of a specific situation. Again, this is why it is imperative to master the use of multiple modalities, which will offer more clues for detecting falsehoods (Yadav et al., 2023).

Another recommendation to address the issues of deception and false content in the metaverse is the adoption of an immediate reactionary approach, which consists of tackling these issues before they become viral. Any intervention that does not take on these problems on time may severely affect individuals or entire communities because it would not only be too late but, at a certain point, nothing could be done about it (Aliverti, 2022). Hence, an immediate reactionary approach operates as a "preventive strike." It is a proactive measure that involves (1) forward-looking predictions to determine risk factors that could negatively impact vulnerable individuals or communities and (2) providing the skills or services necessary to prevent them from being victimized by those risks (Santana Hernández, 2021). A possible challenge is tied to the aforementioned recommender algorithms; they can be the root of the deception and false content problem. To resolve this, users may want to change the algorithms first, instead of immediately tackling fake news (Bojić, 2022).

Finally, with respect to the algorithms themselves, a research approach labeled sentiment analysis could be adopted. The objective is to examine relevant published content in the digital world. More specifically, it could be accomplished through the use of Natural Language Processing, which gives an indication of the author's attitude toward the content. The foundational hypothesis is that any piece of information should ideally reflect a neutral stance vis-à-vis a fact or situation. Any stance skewed toward the left or the right (politically speaking) or one that is either too positive or too negative generally contributes to biased information and, as a result, false information. The goal of sentiment analysis, then, is to provide some computational identification and categorization of opinions expressed through published content. This will allow the analyst to tell whether the author's attitude toward a specific topic was imbued with positive, negative, or neutral tones (Aliverti, 2022).

References

Ahmad, A. R., & Murad, H. R. (2020.): Article e19556). The impact of social media on panic during the COVID-19 pandemic in Iraqi Kurdistan: Online questionnaire study. *Journal of Medical Internet Research, 22*(5). https://doi.org/10.2196/19556

Alevizos, L., Ta, V. T., & Eiza, M. H. (2021.): Article e191). Augmenting zero trust architecture to endpoints using blockchain: A state-of-the-art review. *Security and Privacy, 5*(1). https://doi.org/10.1002/spy2.191

Ali, M., Naeem, F., Kaddoum, G., & Hossain, E. (2022). *Metaverse communications, networking, security, and applications: Research issues, state-of-the-art, and future directions*. arXiv.org. Retrieved on May 24, 2023 from https://arxiv.org/pdf/2212.13993.pdf

Aliverti, M. (2022). The proposed solution: The fake news algorithm project and verification of results. In F. Biondo, G. La Rocca, & V. Trapani (Eds.), *Information disorder: Learning to recognize fake news* (pp. 13–22). Peter Lang.

Allcott, H., & Gentzkow, M. (2017). Social media and fake news in the 2016 election. *Journal of Economic Perspectives, 31*(2), 211–236. https://doi.org/10.1257/jep.31.2.211

Al-Zaman, S. (2021). COVID-19-related social media fake news in India. *Journalism and Media, 2*(1), 100–114. https://doi.org/10.3390/journalmedia2010007

Arafat, S. M., Yasir, A. R., Ahmad, H. R. M., & Kakashekh, H. M. (2021). Perceived impact of social media on panic buying: An online cross-sectional survey in Iraqi Kurdistan. *Frontiers in Public Health, 9*. Article 447.

Bailenson, J. N., Guadagno, R. E., Aharoni, E., Dimov, A., Beall, A. C., & Blascovich, J. (2004). Comparing behavioral and self-report measures of embodied agents' social presence in immersive virtual environments. *Proceedings of the Annual International Workshop on Presence, 7*, 10–21.

Bajwa, A. (2022). Malevolent creativity & the metaverse: How the immersive properties of the Metaverse may facilitate the spread of a mass shooter's culture. *Journal of Intelligence, Conflict, and Warfare, 5*(2), 32–52. https://doi.org/10.21810/jicw.v5i2.5038

Barron, D., Morgan, K., Towell, T., Altemeyer, B., & Swami, V. (2014). Associations between Schizotypy and belief in conspiracist ideation. *Personality and Individual Differences, 70*, 156–159. https://doi.org/10.1016/j.paid.2014.06.040

Berthon, P. R., & Pitt, L. F. (2018). Brands, truthiness and post-fact: Managing brands in a post-rational world. *Journal of Macromarketing, 38*(2), 218–227. https://doi.org/10.1177/0276146718755869

Bibri, S. E. (2022). The social shaping of the Metaverse as an alternative to the imaginaries of data-driven smart cities: A study in science, technology, and society. *Smart Cities, 5*(3), 832–874. https://doi.org/10.3390/smartcities5030043

Blascovich, J. (2002). Social influence within immersive virtual environments. In R. Schroeder (Ed.), *The social life of avatars: Presence and interaction in shared virtual environments* (pp. 127–145). Springer.

Bojić, L. (2022). *Culture organism or techno-feudalism: How growing addictions and artificial intelligence shape contemporary society*. Belgrade: Institute for Philosophy and Social Theory.

Brotherton, R. (2013). Towards a definition of 'conspiracy theory'. *PsyPAG Quarterly, 1*(88), 9–14. https://doi.org/10.53841/bpspag.2013.1.88.9

Brotherton, R., French, C. C., & Pickering, A. D. (2013). Measuring belief in conspiracy theories: The generic conspiracist beliefs scale. *Frontiers in Psychology, 4*, Article 279. https://doi.org/10.3389/fpsyg.2013.00279

Brown, J. G., Bailenson, J. N., & Hancock, J. (2023). Misinformation in virtual reality. *Journal of Online Trust and Safety, 1*(5), 1–30. https://doi.org/10.54501/jots.v1i5.120

Buller, D. B., & Burgoon, J. K. (1996). Interpersonal deception theory. *Communication Theory, 6*(3), 203–242. https://doi.org/10.1111/j.1468-2885.1996.tb00127.x

Campbell, C., Plangger, K., Sands, S., & Kietzmann, J. (2022). Preparing for an era of deepfakes and AI-generated ads: A framework for understanding responses to manipulated advertising. *Journal of Advertising, 51*(1), 22–38. https://doi.org/10.1080/00913367.2021.1909515

Chan, T. K. H., Cheung, C. M. K., & Wong, R. Y. M. (2019). Cyberbullying on social networking sites: The crime opportunity and affordance perspectives. *Journal of Management Information Systems, 36*(2), 574–609. https://doi.org/10.1080/07421222.2019.1599500

Cinelli, M., Brugnoli, E., Schmidt, A. L., Zollo, F., Quattrociocchi, W., & Scala, A. (2020.): Article e0229129). Selective exposure shapes the Facebook news diet. *PLoS One, 15*(3). https://doi.org/10.1371/journal.pone.0229129

Cohen, S. (1972). *Folk devils and moral panics*. Routledge.

Crossman, A. (2019). *A sociological understanding of moral panic*. ThoughtCo. Retrieved on May 20, 2023 from https://www.thoughtco.com/moral-panic-3026420

Ekman, P. (1992). *Telling Lies*. W. W. Norton.

Epley, N., & Gilovich, T. (2016). The mechanics of motivated reasoning. *Journal of Economic Perspectives, 30*(3), 133–140. https://doi.org/10.1257/jep.30.3.133

Fernandez, C. B., & Hui, P. (2022). Life, the metaverse and everything: An overview of privacy, ethics, and governance in metaverse. *IEEE International Conference on Distributed Computing Systems Workshops (ICDCSW), 42*. https://doi.org/10.1109/ICDCSW56584.2022.00058

Fraga-Lamas, P., & Fernández-Caramés, T. M. (2020). Fake news, disinformation, and deepfakes: Leveraging distributed ledger technologies and blockchain to combat digital deception and counterfeit reality. *IT Professional, 22*(2), 53–59. https://doi.org/10.1109/MITP.2020.2977589

Freeman, D., & Bentall, R. P. (2017). The concomitants of conspiracy concerns. *Social Psychiatry and Psychiatric Epidemiology, 52*(5), 595–604. https://doi.org/10.1007/s00127-017-1354-4

González-Bailón, S., & Wang, N. (2016). Networked discontent: The anatomy of protest campaigns in social media. *Social Networks, 44*(1), 95–104. https://doi.org/10.1016/j.socnet.2015.07.003

Granovetter, M. S. (1973). The strength of weak ties. *American Journal of Sociology, 78*(6), 1360–1380. https://doi.org/10.1086/225469

Guadagno, R. E., Blascovich, J., Bailenson, J. N., & Mccall, C. (2007). Virtual humans and persuasion: The effects of agency and behavioral realism. *Media Psychology, 10*(1), 1–22. https://doi.org/10.1080/15213260701300865

Guo, H., Yantao, Y., Xiang, T., Li, H., & Zhang, D. (2017). The availability of wearable-device-based physical data for the measurement of construction workers' psychological status on site:

From the perspective of safety management. *Automation in Construction, 82*, 207–217. https://doi.org/10.1016/j.autcon.2017.06.001

Gupta, A., Khan, H. U., Nazir, S., Shafiq, M., & Shabaz, M. (2023). Metaverse security: Issues, challenges and a viable ZTA model. *Electronics, 12*(2), Article 391. https://doi.org/10.3390/electronics12020391

Hadnagy, C. (2018). *Social engineering: The science of human hacking* (2nd ed.). Wiley.

Han, D.-I. D., Bergs, Y., & Moorhouse, N. (2022). Virtual reality consumer experience escapes: Preparing for the metaverse. *Virtual Reality, 26*(4), 1–16. https://doi.org/10.1007/s10055-022-00641-7

He, Y., Huang, D., Chen, L., Ni, Y., & Ma, X. (2022). A survey on zero trust architecture: Challenges and future trends. *Wireless Communications and Mobile Computing, 2022.*): Article 6476274. https://doi.org/10.1155/2022/6476274

Hossain, M. A., Chowdhury, M. H., Pappas, I. O., Metri, B., Hughes, L., & Dwivedi, Y. K. (2022). Fake news on Facebook and their impact on supply chain disruption during COVID-19. *Annals of Operations Research, 327*(5), 1–29. https://doi.org/10.1007/s10479-022-05124-1

Hotez, P. J. (2016.): Article e1002153). Texas and its measles epidemics. *PLoS Medicine, 13*(10). https://doi.org/10.1371/journal.pmed.1002153

Huq, N., Reyes, R., Lin, P., & Swimmer, M. (2022). *Metaverse or metaworse: cybersecurity threats against the internet of experiences* (pp. 1–24). Trend Micro. Retrieved on May 30, 2023 from http://documents.trendmicro.com/assets/white_papers/wp-metaverse-or-metaworse-cybersecurity-threats-against-the-internet-of-experiences.pdf

Ibarra, H., Kilduff, M., & Tsai, W. (2005). Zooming in and out: Connecting individuals and collectivities at the frontiers of organizational network research. *Organization Science, 16*(4), 359–371. https://doi.org/10.1287/orsc.1050.0129

Jia, C., & Johnson, T. J. (2021). Source credibility matters: Does automated journalism inspire selective exposure? *International Journal of Communication, 15*, 3760–3781.

Jung, S. H., & Jeon, I.-O. (2022). A study on the components of the metaverse ecosystem. *Journal of Digital Convergence, 20*(2), 163–174.

Kadushin, C. (2012). *Understanding social networks: Theories, concepts, and findings*. Oxford University Press.

Kalniņš, R., Puriņš, J., & Alksnis, G. (2017). Security evaluation of wireless network access points. *Applied Computer Systems, 21*(1), 38–45. https://doi.org/10.1515/acss-2017-0005

Kantharaju, R. B., De Franco, D., Pease, A., & Pelachaud, C. (2018). Is two better than one? Effects of multiple agents on user persuasion. *Proceedings of the International Conference on Intelligent Virtual Agents, 18*, 255–262. https://doi.org/10.1145/3267851.3267890

Kim, A., & Dennis, A. R. (2019). Says who? The effects of presentation format and source rating on fake news in social media. *MIS Quarterly, 43*(3), 1025–1039. https://doi.org/10.2139/ssrn.2987866

Kumar, S., Talwar, S., Krishnan, S., Kaur, P., & Dhir, A. (2021.): Article 102668). Purchasing natural personal care products in the era of fake news? The moderation effect of brand trust. *Journal of Retailing and Consumer Services, 63*(1). https://doi.org/10.1016/j.jretconser.2021.102668

Larson, J. M. (2017.): Article 14). The weakness of weak ties for novel information diffusion. *Applied Network Science, 2*(1). https://doi.org/10.1007/s41109-017-0034-3

Lazarsfeld, P. F., Berelson, B., & Gaudet, H. (1968). *The people's choice: How the voter makes up his mind in a presidential campaign*. Columbia University Press.

Liang, G., Weller, S. R., Zhao, J., Luo, F., & Dong, Z. Y. (2017). The 2015 Ukraine blackout: Implications for false data injection attacks. *IEEE Transactions on Power Systems, 32*(4), 3317–3318. https://doi.org/10.1109/TPWRS.2016.2631891

Mason, L. E., Krutka, D. G., & Stoddard, J. (2018). Media literacy, democracy, and the challenge of fake news. *Journal of Media Literacy Education, 10*(2), 1–10. https://doi.org/10.23860/JMLE-2018-10-2-1

Matusitz, J., Madrazo, A., & Udani, C. (2019). *Online jihadist magazines to promote the caliphate: Communicative perspectives*. Peter Lang.

Mazarr, M. J., Bauer, R., Casey, A., Heintz, S., & Matthews, L. J. (2019). *The emerging risk of virtual societal warfare: Social manipulation in a changing information environment*. RAND Corporation.

McLennan, R. (1970). Selective exposure to information. *Management Decision, 4*(2), 23. https://doi.org/10.1108/eb000930

Melki, J., Kozman, C., Mellado, C., Elórtegui, C., So, C. Y. K., Movahedian, M., Salim, S. K., & Farhat, S. (2023). Selective exposure during uprisings: A comparative study of news uses in Chile, Hong Kong, Iran, Iraq, and Lebanon. *International Communication Gazette, 85*(7), Article 174804852211445. https://doi.org/10.1177/17480485221144582

Mittal, V., Huppertz, J. W., & Khare, A. (2008). Customer complaining: The role of tie strength and information control. *Journal of Retailing, 84*(2), 195–204. https://doi.org/10.1016/j.jretai.2008.01.006

Mustafaraj, E., & Metaxas, P. T. (2010). From obscurity to prominence in minutes: Political speech and real-time search. *Proceedings of the Web Science Conference: Extending the Frontiers of Society On-Line*. (April 26, 2010). Retrieved on June 5, 2023 from https://repository.wellesley.edu/object/ir122

Nguyen, C. T., Hoang, D. T., Nguyen, D. N., & Dutkiewicz, E. (2022). MetaChain: A novel blockchain-based framework for metaverse applications. *IEEE Vehicular Technology Conference, 95*. https://doi.org/10.1109/VTC2022-Spring54318.2022.9860983

Nunes, J. C., Ordanini, A., & Giambastiani, G. (2021). The concept of authenticity: What it means to consumers. *Journal of Marketing, 85*(4), 1–20. https://doi.org/10.1177/0022242921997081

Obadă, R. (2019). Sharing Fake News about brands on social media: A new conceptual model based on flow theory. *Argumentum. Journal of the Seminar of Discursive Logic, Argumentation Theory and Rhetoric, 17*(2), 144–166.

Pajuste, T. (2022). *Specific threats to human rights protection from the digital reality*. Tallinn University.

Park, A., Montecchi, M., Feng, C. M., Plangger, K., & Pitt, L. (2020). Understanding fake news: A bibliographic perspective. *Defence Strategic Communications, 8*, 141–172. https://doi.org/10.30966/2018.RIGA.8.4

Pee, L. G. (2018). Affordances for sharing domain-specific and complex knowledge on enterprise social media. *International Journal of Information Management, 43*, 25–37. https://doi.org/10.1016/j.ijinfomgt.2018.05.006

Petratos, P. N. (2021). Misinformation, disinformation, and fake news: Cyber risks to business. *Business Horizon, 64*(6), 763–774. https://doi.org/10.1016/j.bushor.2021.07.012

Plangger, K., & Campbell, C. L. (2022). Managing in an era of falsity: Falsity from the metaverse to fake news to fake endorsement to synthetic influence to false agendas. *Business Horizons, 65*(6), 10–21. https://doi.org/10.1016/j.bushor.2022.08.003

Repez, F., & Popescu, M.-M. (2020). Social media and the threats against human security against human security deepfake and fake news. *Romanian Military Thinking, 4*, 44–55.

Robinson, O., Coleman, A., & Sardarizadeh, S. (2019). *A report on anti-disinformation initiatives*. Oxford Technology and Elections Commission.

Rohloff, A., & Wright, S. (2010). Moral panic and social theory: Beyond the heuristic. *Current Sociology, 58*(3), 403–419. https://doi.org/10.1177/0011392110364039

Salahdine, F., & Kaabouch, N. (2019.): Article 89). Social engineering attacks: A survey. *Future Internet, 11*(4). https://doi.org/10.3390/fi11040089

Santana Hernández, J. D. (2021). Between reaction and prevention: A systematic review of a social intervention dilemma for social work (2000–2019). *Cuadernos de Trabajo Social, 34*(1), 103–114.

Scheufele, D. A., & Krause, N. M. (2019). Science audiences, misinformation, and fake news. *Proceedings of the National Academy of Sciences, 116*(16), 7662–7669. https://doi.org/10.1073/pnas.1805871115

Shao, C., Ciampaglia, G. L., Varol, O., Yang, K.-C., Flammini, A., & Menczer, F. (2018.): Article 4787). The spread of low-credibility content by social bots. *Nature Communications, 9*(1). https://doi.org/10.1038/s41467-018-06930-7

Shu, J., Shen, X., Liu, H., Yi, B., & Zhang, Z. (2018). A content-based recommendation algorithm for learning resources. *Multimedia Systems, 24*(1), 163–173. https://doi.org/10.1007/s00530-017-0539-8

Smith, M. (2017). How Facebook, fake news and personalised ads could swing the 2017 election—And what you can do about it. *The Mirror*. (May 8, 2017): A1. Retrieved on February 11, 2023 from https://www.mirror.co.uk/news/politics/how-facebook-fake-news-personalised-10382585

Stroud, N. J. (2018). Selective exposure theories. In K. Kenski & K. H. Jamieson (Eds.), *The Oxford handbook of political communication*. Oxford University Press.

Szűts, Z. (2021). Online identities in politics. Technological and content-based approach. *Acta Universitatis Sapientiae, Communicatio, 8*, 60–69. https://doi.org/10.2478/auscom-2021-0005

Theocharis, Y., Cardenal, A., Jin, S., Aalberg, T., Hopmann, D. N., Strömbäck, J., Castro, L., Esser, F., Van Aelst, P., de Vreese, C., Corbu, N., Koc-Michalska, K., Matthes, J., Schemer, C., Sheafer, T., Splendore, S., Stanyer, J., Stępińska, A., & Štětka, V. (2021). Does the platform matter? Social media and COVID-19 conspiracy theory beliefs in 17 countries. *New Media & Society, 25*(12), 3412–3437. https://doi.org/10.1177/14614448211045666

Treem, J. W., & Leonardi, P. M. (2013). Social media use in organizations: Exploring the affordances of visibility, editability, persistence, and association. *Annals of the International Communication Association, 36*(1), 143–189. https://doi.org/10.1080/23808985.2013.11679130

Vosoughi, S., Roy, D., & Aral, S. (2018). The spread of true and false news online. *Science, 359*(6380), 1146–1151. https://doi.org/10.1126/science.aap9559

Walker, J. (2014). *The United States of paranoia: A conspiracy theory*. Harper Perennial.

Walsh, J. P. (2020). Social media and moral panics: Assessing the effects of technological change on societal reaction. *International Journal of Cultural Studies, 23*(6), 840–859. https://doi.org/10.1177/1367877920912257

Weber, M. S., & Haseki, M. (2021). Social media affordances to engage clients during the sales process: Sequential versus multiplex media use. *International Journal of Electronic Commerce, 25*(1), 73–98. https://doi.org/10.1080/10864415.2021.1846855

Weeks, B. E., & de Ziga, H. G. (2019). What's next? Six observations for the future of political misinformation research. *American Behavioral Scientist, 65*(2), 277–289. https://doi.org/10.1177/0002764219878236

White (Ed.). (2002). The value of conspiracy theory. *American Literary History, 14*(1), 1–31. https://doi.org/10.1093/alh/14.1.1

Wong, C. (2021). Official Chinese social media accounts forced to delete posts mocking India's Covid-19 death toll. *South China Morning Post*. (May 3, 2021): A1. Retrieved on May 20, 2023 from https://www.scmp.com/news/china/diplomacy/article/3132109/official-chinese-social-media-accounts-forced-delete-posts

Woolley, S. (2020). *The reality game: How the next wave of technology will break the truth*. Public Affairs.

Yadav, A., Gaba, S., Khan, H., Budhiraja, I., Singh, A., & Singh, K. K. (2023). *ETMA: Efficient transformer based multilevel attention framework for multimodal fake news detection*. arXiv. https://doi.org/10.48550/arXiv.2206.07331

Zhang, X., & Ghorbani, A. A. (2020.): Article 102025). An overview of online fake news: Characterization, detection, and discussion. *Information Processing & Management, 57*(2). https://doi.org/10.1016/j.ipm.2019.03.004

Zheng, J. M., Chan, K. W., & Gibson, I. (1998). Virtual reality. *IEEE Potentials, 17*(2), 20–23. https://doi.org/10.1109/45.666641

Chapter 9
Other Problems: Privacy Concerns, Surveillance, and Hate Speech

When the pioneer of the World Wide Web, Tim Berners-Lee (2018), argues that the web is in peril because of false information, dubious advertising, greedy marketing practices, the difficulty in hiding one's personal information, and the weaponization of the virtual by big-tech corporations, then theorists and practitioners alike ought to be concerned about the fate of metaverse users. By weaponization, Berners-Lee alludes to the misapplication of communication and the exploitation of data for political, ideological, financial, or personal gain. An oft-cited example is the 2018 Facebook data breach—i.e., users' personal information that was collected without their consent by British consulting firm Cambridge Analytica—to target and manipulate helpless citizens (Venturini & Rogers, 2019). Among a great number of other issues, the algorithms in the metaverse are in full swing. Aware of the unlimited forms of user's interactions and activities in the virtual environment, these algorithms seek to stockpile sensitive personal data about users' sexual orientation, race, gender, academic achievement, and even past trauma—thereby mining and refining the entire apparatus of the digital economy. These massive volumes of personal information can be capitalized on target users with custom-built information to sway them into engaging in particular courses of action (Cadwalladr & Graham-Harrison, 2018). This chapter sets out to examine three core problems in the metaverse: privacy concerns, surveillance, and hate speech.

Privacy Concerns

Distress about lack of privacy in the metaverse encompasses problems such as unwarranted data sharing and data breaches (among others). Sensitive data can be misused when information is taken or shared without a legitimate or agreed-upon purpose, as would be the case in marketing or government surveillance. Hacking and identity theft are also major reasons for anxiety (Curtis & Brolan, 2023). Ever

J. Matusitz, J. Dacas, *Communication in the Metaverse*, Signals and Communication Technology, https://doi.org/10.1007/978-3-031-63485-7_9

since humankind moved into digital and augmented realities for both business and pleasure, the volume of user data has been commensurate with that; users' behaviors and emotional states continue to be exploited in innovative ways without their explicit permission. Only a small percentage of users possess the intuitive interface to observe and control their privacy settings in a safe way. In the early stages of widely available digital and augmented realities, new privacy equipment was thought to be advantageous because it would replace headphones and goggles and, by the same token, allow participants to adjust their privacy options. However, privacy options differ from audio and visual settings because they involve a different context not immediately tied to the senses, but they also encompass a much larger sphere of physical, physiological, and emotional layers. Although settings impacting the users' experiences can be generally managed in visual, audio, and haptics terms, data produced by the users may also incorporate what users rarely think about: their heart rate, body movements, facial expressions, physical location, and so forth (Yao et al., 2021).

How Could Attacks Technically Work?

Privacy is one of the greatest threats associated with the metaverse. Virtual reality (VR) technologies can observe what an individual does, putting his or her privacy at risk. In contrast to other forms of technology—such as those used in the three-dimensional world—the metaverse collects a great amount of information about the user. Wei et al. (2022) explained how the development of an augmented reality (AR) application employs computer vision and basic input data to wage an instantaneous attack on an attacker's portable digital device. To conduct an attack, this procedure emulates each user's unique typing behavior on a smartphone without altering the actual content on the device. To be more precise, it installs software or special hardware without the user's knowledge. The attacker will operate from their own device to run this app, positioning in a way that the camera can have a plain view of the user's hands as they handle their phone. By using this strategy, the attacker can mimic the victim's behavior by giving instructions via the camera stream. To generate an AR prototype, they construct a basic physical model. An attacker can also utilize audio beeps to time their taps and affix a transparent film with spatial hints to the target's smartphone (Pooyandeh et al., 2022).

It is facile to postulate that the metaverse is the ideal alternative to the physical world. This assumption is predicated on the belief in a "universal human existence" by existing digital developments. In the real world of present-day society, an array of artifacts (i.e., information credentials) that can speak to the "I"—centered on our bodies—percolate through all layers of social life. They are proof of one's own existence, especially when identity cards, credit cards, work permits, and portable digital devices become major targets of attacks. With the evolution of big data technologies, details of our personal identity begin to merge in various domains such as healthcare, household, financial, and others. Hence, as the consolidation and interoperability of identity information fragments progressively surges, a universal, omnipresent e-Existence around each of us emerges from it. In this manner, inhabitants of the metaverse are not only physically present, but they also express the "I" in the powerful integration of data and information, and this process is happening fast (Hu & Liu, 2022).

The strength of communication and confidentiality in the metaverse is not immutable. As we have seen, security issues, privacy threats, data leakage, and the like can occur because of these characteristics. In some cases, invaders may expose confidential documents and meeting memos to wreak significant economic havoc to companies. This is why "access control" is necessary to guarantee that only authenticated and authorized users can access vital information. This can be achieved through attribute-based access control and role-based access control. By extension, in the event that documents are leaked, they must be made useless through encryption. By and large, symmetric-key cryptography, such as the state-of-the-art encryption standard, is included in encryption algorithms. To ensure that meeting contents are not leaked, secure communication through transport layer security and a digital private network are also needed (Kang et al., 2023).

In the Healthcare-Based Metaverse

Data management privacy and user safety issues in the metaverse stem from challenges facing VR platforms and the video game industry as a whole (Needleman, 2021). Within the healthcare field, in particular, privacy is of utmost importance. Patient records must be stored within authorized spaces in order to protect both patients and providers. In a healthcare-based metaverse, information can be obtained by an invader because storage spaces always have a vulnerable spot. A Man-in-the-Middle (MIM) attacker—i.e., a perpetrator who sneaks in the middle of communication to eavesdrop or impersonate (Al-Shareeda & Manickam, 2022)—can view the flow of information between the body sensor network and the metaverse. In Falchuk, Loeb, and Neff's (2018) publication, the main line of reasoning is that patients' pervasive data collection can be easily appropriated by a MIM attacker who is spectating the user in the metaverse. The essence of MIM attacks is for intelligence, surveillance, and infiltration; they operate on the principle of silently obtaining sensitive data. In an environment replete with naïve users, such as the metaverse, identifying the likely presence of an attack is not evident. Therefore, prevention methods must be adopted (Mejia & Rawat, 2022).

Metaverse applications in healthcare are used the most in the telemedicine and psychological counseling domains. Whereas they can conquer the limitations of space, distance, and time and, thereby, deliver higher-quality medical services, the data in those healthcare applications can automatically reveal the users' identity and their disease information. This explains, in part, why the applications for healthcare are typified as "sensitivity" and "confidentiality." "Data leakage" and "excessive data collection" can happen in the process. As such, sensitive medical data can be exposed, which may make the user inadvertently discriminated against or put his or her life in danger. In addition, perpetrators can make financial gains by intimidating parties into disclosing e-health information. Marketers can also gather unwarranted medical data in their marketing strategies (Kang et al., 2023).

Six Categories of Privacy Concerns

In this section, six categories of privacy concerns illustrate how dangerous the metaverse can be: (1) context collapse, (2) privacy leakage during data transmission, (3) privacy leakage during data processing, (4) compromised or rogue end devices, (5) unlawful monetization of data, and (6) stalking.

Context Collapse

Real-time activities, online behaviors, and interactions in the metaverse may give rise to a pervasive problem called context collapse (Marwick & Boyd, 2011). Context collapse is a phenomenon whereby a participant's online communications are displayed to unwanted multiple audiences. On traditional social media sites, a Facebook post can be shared as a tweet by a person who was not the original audience of that post. The metaverse was not designed to record and keep track of all activities. However, it is now relatively easy to make others "aware" of our online communications. In fact, metaverse platforms will divulge a lot of information about users even when they move between different metaverse spaces. This is why context occasionally collapses in the metaverse, but in manners different from those happening on social media. The metaverse does not rely on digital traces; in other words, users have the ability to precipitously show up into a scene or performance (such as a digital game or a shopping experience). It should be mentioned that metaverse administrators can set boundaries as to whom enters a scene or performance. For example, it is expected that sex-related metaverse experiences are taking measures to prevent minors from joining such VR platforms—indeed, one should never be too naïve in believing that a significant chunk of metaverse operations will not involve pornography (Ahmed et al., 2016).

Privacy Leakage During Data Transmission

Metaverse operations involve the sharing of immense amounts of sensitive personal data that is gathered via portal digital devices (or head-mounted displays) and transmitted by means of wireless and wired communications. Unauthorized parties have accessed such confidential information (Ometov et al., 2017). Notwithstanding the fact that the information, if truth be told, is transferred confidentially and communication is encrypted, as described earlier, attackers find ways to intercept it or eavesdrop it on particular channels (Wasserkrug et al., 2008); they also wage various attacks to identify users' locations (Wei et al., 2021). Exploiting people's vulnerabilities to provide data for big-tech corporations (digital footprint) while enjoying life in a virtual home (e.g., on social media or in the metaverse) `can be equated with modern slavery or as some type of techno-feudalism (Waters, 2021).

Privacy Leakage During Data Processing

Organizations and healthcare institutions find it indispensable to store and gather massive volumes of data from customers or patients to render day-to-day operations and administrative functions in virtual environments smoother. Yet, such data processing can cause leakage (Li et al., 2022). More specifically, perpetrators have been known to obtain users' private details by delving into diverse processing outcomes (e.g., synthetic avatars) in virtual settings such as *Fortnite* and Roblox (Ali et al., 2022). In addition, this type of data processing presents a considerable danger because the servers, assets, and users are usually not in the same "locations." The cloud-distributed character of the metaverse means that data is processed and stored in various locations that are disconnected from the user (Huq et al., 2022).

Compromised or Rogue End Devices

The metaverse involves, inter alia, the placement of multiple wearable sensors on users' bodies and surroundings to allow avatars to mirror facial expressions, hand gestures, and natural-looking eye contact—all of which happens in real-time. A predominant threat is that the wearable devices may have already recorded who you are and the ways in which you express, feel, act, and talk (Ali et al., 2022). A great deal of data processing occurs at the user endpoints (e.g., AR/VR headsets) that are becoming highly influential computing platforms. User identity data (e.g., life science statistics), location data, professional accomplishments, payment information, avatar information, and so forth are stored locally in user endpoints, thereby making them an ideal target for financial sharks. AR applications that have iris tracking make it easier for them to steal our iris patterns. As reported by a study published in *Nature*, users' body motions are easily identifiable (Miller et al., 2020). In the metaverse, compromised or rogue wearable end devices have become an avenue for malware invasions and data leakages. This problem will be exacerbated when such devices grow in popularity (Shang et al., 2022). Under the surveillance of those compromised or rogue end devices, the avatars themselves will be used as a data collection source in the metaverse, making user privacy infringement a daily habit (Ali et al., 2022).

Unlawful Monetization of Data

It is clear by now that a major preoccupation in the metaverse is data privacy. People's information can be appropriated by invaders through illicit contracts, malware, and spyware, in concert with more phishing attacks. Similarly, AR and VR devices have already been proven as possessing significant loopholes. Although the metaverse has been touted as leading to a virtual copy of real life, it will allow malevolent ones to obtain users' personal information, including body-related information, biometric data, health information, consumer preferences, and so on. The lack of proper and strict documentation of legal policies enables such malevolent ones to make illegal monetization of data, unbeknownst to private users or companies (Salau et al., 2022). It can occur, for example, in a situation where various interested persons are led to believe that X, Y, Z applications are free to use at the basic level—yet, there would be a fee for each upgrade. Since we know that there is no such thing as a free lunch, hackers will indirectly benefit from many aspects of meta spaces, collect huge amounts of money, or illegally monetize collected users' personal data. Despite the wide availability of open source metaverse applications that users can host, other "parties" can still collect and monetize user data without anybody's permission (Huq et al., 2022).

Stalking

Stalking is certainly an issue. Ill-intentioned individuals can stalk a victim through an invisible persona (frequently, it is an invisible avatar). The victim will not be aware that he or she is being tailed, monitored, or eavesdropped on. Invisible avatars rely on a user account with a higher privilege, which can be obtained by taking advantage of some vulnerability in the metaverse publisher's code. Thus, one's real identity is highly correlated with one's virtual identity in the metaverse. Even though users often have multiple identities, such as in social media platforms, with no real connections to their real selves, the platforms themselves can still be exploited to accurately identify the real person behind any digital persona; this can be done with parameters such as body motion data and user behavior. Online users should learn how to use customizable and understandable options to modify the algorithms that manage their data. These options ought to be laid out clearly in a standardized fashion. At any time, the user should be able to delete their whole digital footprint (to avoid being stalked). The digital footprint should be made available to all through special open-access technology, such as a blockchain (Bojić, 2022a). Naturally, on account of stalking and privacy issues, there is too much identity confusion in these spaces. Individuals tend to have multiple identities in the same space and across metaverse platforms, engendering trust issues.

Surveillance Capitalism

In the context of this chapter, surveillance capitalism refers to a takeover from above, not an overthrow of the state but rather a seizure of the citizens' sovereignty. It is a radical, perilous shift toward a process of undemocratizing free and open societies. Platforms that are ideological or politically biased act like big-tech corporations that abuse their power, leading to dire circumstances of disciplinary surveillance through a technocracy. Put simply, digital technologies are exploited for capture and control (Gray, 2019). The pinnacle of surveillance capitalism is an extreme form of social order that is totalitarian, which clashes with the image presented by big-tech companies that the metaverse heralds an age of transparency, accountability, and freedom (Zuboff, 2015).

At the Heart of the Metaverse

Surveillance is at the heart of the metaverse. It is effectuated through a process of platformization of its data infrastructures and governance frameworks (Bibri, 2022). The assurance of a better life in the metaverse can turn out to be an apocalyptic nightmare for certain groups of users in different parts of the world, as their options for creative adventures or explorations would be greatly diminished (Crutzen, 2005). Bibri and Allam (2022) looked at the forms, activities, and ethics of the metaverse in the context of data-driven smart cities, with a special focus on privacy, dataveillance, and geosurveillance, among others. The authors discovered that the consequences that the metaverse will have on the users' daily experiences can be disastrous. If mishandled, the metaverse can cause more harm than good to users because of the misuse of it—particularly with respect to the datafication, algorithmization, and platformization underpinning the global apparatus of the computer mediation upon which surveillance capitalism relies.

Wiretapping Issues

It should come as no surprise that wiretapping is a major issue in the metaverse. Every move or word by a user can be subject to unlimited surveillance. This problem can grow even more extreme. Unless drastic laws are passed and implemented, it can be a free-for-all. Metaverse operators could have an unparalleled view of user actions. In essence, when operating in a metaspace, any move or word can be wiretapped. This situation is akin to present-day social media, where applications are seemingly free, but the users may be exploited with their personal data. Metaverse operators can obtain vast amounts of user data and capitalize on it (Huq et al., 2022). Wiretapping issues have been particularly flagged as dangerous during training and recruitment sessions. Whereas employers may try to tap into the telemetry that metaverse platforms provide—i.e., it could record the productivity of a person or an entire team—employees might strongly disagree and fight back. In a survey including remote and hybrid workers (and published through an ExpressVPN report), almost 60% expressed anxiety about their employer monitoring them. Over 50% admitted that they would resign from their positions if their employer put surveillance measures into practice (cited in Wiggers, 2022).

Monitoring Public Opinion

There is a wide range of possibilities for monitoring public opinion in the metaverse. Even a user's past public opinions can leave a digital footprint after being concealed. The reason is that the metaverse has a digital memoir of that user's alternate life online. More specifically, personal data (including opinions) that are centrally stored in cloud servers or edge devices can be retrieved by clever administrators. Profiling, behavioral detection (through patterns), and social engineering do not only constitute a serious violation of an individual's privacy, but it can also create situations that favor alternative opinions (opinions that will progressively condition users into believing a certain ideology). In this sense, it is a form of indoctrination. This surveillance practice is not confined to only one end of the process of virtual interactions. It has also been observed in VR devices that get tampered with, along with similar devices that get maneuvered to track physical movements. The metaverse itself can become a passive enabler of surveillance, data breaches, and privacy/confidentiality violations because of a faulty algorithm or software application. This can lead to users' problems with little legal recourse—i.e., an allegation of techno-negligence is not necessarily simple to prove. Surveillance can also happen at various stages during an end-to-end data transmission process, beginning at the data collection stage and ending with the data transfer step. Encryption and distributed transmission techniques can greatly alleviate such threats, but they do not automatically eliminate it (Sethi, 2022).

Monitoring Body Language

In addition to tracking public opinion, the metaverse can monitor how users act, move, do, what they look at, and how they look at things. Tracking entails real-time monitoring of users' gaze, their gait, their speed, their stoppage time, and their attempt to pass locations—among many other things. Metaverse platforms are even able to surveil manual reach; that is, determining when users reach for objects (both real and virtual) and measuring the time they take to hold such objects. AR has been ascertained as capable of monitoring users' grab for real objects in the real world. An illustration would be a situation in which users go shopping in physical stores that are augmented with virtual elements (Rosenberg, 2022). This may appear ultra-modern, but real-time tracking of human body language vis-à-vis real objects was already possible through interactive AR systems at the Air Force Research Laboratory in 1992 (Rosenberg, 2021).

Immersive platforms have an exceptionally high aptitude at detecting bodily movements, facial expressions, and vocal inflections, whereas intelligent algorithms can exploit such data to anticipate each individual's real-time emotional state (Ivanova & Borzunov, 2020). The consequence is that metaverse administrators can not only know how their users act and interact with their bodies, but they can also know how they emotionally react as they navigate real and digital spaces. More specifically, their responses can be profiled at much deeper levels than had been possible within older media formats. Of course, the concern does not only pertain to the fact that these personal parameters can be monitored in real-time; advertisers and other relevant third parties can capitalize on such invasive data to maneuver the consumers' desires and needs more efficiently than ever before (Rosenberg, 2022).

Case Study: Metaverse Authoritarianism in China

The metaverse complex in China is expected to grow by almost 40% annually, reaching over US$ 100 billion in value over the next few years (an estimate by the consulting firm Tech Insight 360). The country's expenditures on the metaverse will increase by 500%, going from US$ 76 billion in 2022 to a whopping US$ 457 billion by 2030 (Williams, 2023). From a more theoretical perspective, the main sources of power for big-tech corporations are ideology, truth-bending, and alternative culture and morals. It is imperative to identify those with rule-making power in the metaverse. The recommender systems that expose metaverse users to content will probably maintain their big-tech monopolies in the Western world. In China, things operate differently (Bojić, 2022b).

The Chinese Communist Party (CCP)

In China, the growth of a massive technology does not necessarily happen organically in a bottom-up, consumer-driven fashion. This case study of China is evidence that technologies can scale exponentially without consumer buy-in. Here, it is caused by a top-down rollout of technology (Parasol, 2022). It is important to acknowledge that similar phenomena occur beyond China. The ruling Chinese Communist Party (CCP) heavily relies on VR tools to evaluate and consolidate party loyalty. For example, in the Shandong province of eastern China, party members are obligated to take part in political loyalty tests that use VR. As reported by a popular industry blog called VR Focus, the "Test of Dangxing," or "Test of Party Spirit," was mandatory for party members in Qingyang as well (a prefecture-level city in the Gansu province in northwestern China). They had to wear VR headsets and join a virtual meeting, in which leading figures of the CCP quizzed them on a wide spectrum of subjects, ranging from Communist Party theory to the pioneering roles of the CCP. More importantly, party members were questioned about their own family and social lives (Hills-Duty, 2018).

One might wonder what the CCP's reasons are for using the metaverse experience, in lieu of a phone or video conference call. In either case, the CCP would have asked the same questions. To begin, VR is more immersive and, hence, more physiologically cogent. At the same time, it is also intimidating because party members' learned threat responses have been conditioned by real, not virtual, environments for many years (Rosén et al., 2019). The ramifications of bad performance or dishonest behavior feel more real in a deeply immersive environment. According to the élite that administered the VR experiences, the test results were helpful in informing the CCP and identifying those people who have the best qualities for promotion within the CCP. Thus, the metaverse can be a supreme medium for manipulation of the human mind—not only through misinformation and disinformation but also through mind control (Hills-Duty, 2018). Considering that China exported over 80% of global shipments for VR headsets in 2019, these developments are likely to benefit the CCP (Baptista, 2019).

Compliance and Punishment

Compliance is expected, and punishment is the result of noncompliance. Although enforcement in the United States and the European Union is less stringent or severe, it is certainly a deterrent for companies and individuals in China. As a single-party authoritarian system, the CCP has the edge in legislative flexibility—especially since these measures were put in place before the United States or the European Union. China's regulations do not merely relate to deepfakes but also generated text, visual enhancements, and virtual scenes. This signifies that the CCP is thinking about a future that transcends current digital applications; it would be more about an emerging technology that could safeguard the viability of its régime—by anticipating future developments in VR environments. It also demonstrates how the CCP is more reliant on monopolizing this type of media than the more liberal US and European governments. As the Chinese state exerts much more influence over what digital platforms its citizens can access, the monopolizing option seems like a more realistic strategy than it would be for US or European powers. These profound differences in governance structures have fashioned China's enforcement abilities, which explains why it is so easy for the CCP to enforce compliance and punishment. As a totalitarian law-abiding régime, where law equates political control, China has a much more powerful enforcement mechanism than the United States and the European Union, where the rule of law is designed to be above politics (Hine & Floridi, 2022).

Apropos this radical, but also efficient, information revolution, it can be deduced that this case study of China is not merely about digital politics and power, but it is also about the reality that the information revolution is not objective with respect to its magnitude, scope, or the rapidity with which it can be implemented. Rather, it is created, processed, and evaluated according to a specific ideological system (whether the Chinese people are cognizant of this reality or not). In this manner, the metaverse is exploited to fashion new forms of power relations and politics, which ultimately benefits only the CCP and its subservient élite (cited in Egliston & Carter, 2021). From this standpoint, identifying metaverse-related harms and contemplating how scholars and practitioners might productively, and more equitably, tackle them would be instrumental in bettering humankind (Couldry & Mejias, 2019).

Possible Solution: Virtual Realist Governance

Most organizations and corporations have some presence in the metaverse, with some companies being 100% involved in it. It is imperative to ensure that metaverse headquarters—the leading NFT Alpha Community—be safe and that data be protected. A certain degree of cyber-physical security is necessary to protect such headquarters to eschew sabotage and attacks (Huq et al., 2022). This is about physical security, but a broader solution is virtual realist governance. Virtual realist governance rests on the long-established principle of virtual realism and David Chalmer's recent theorizing of Reality+. Essentially, the virtual should be taken to be real, implying that experiences of virtual objects and what happens in the three-dimensional world should be considered as worthy and meaningful (McStay, 2023b). The hypothesis of virtual realism was initially proposed by Heim (1998) and developed further by Chalmers (2022). According to the latter, virtual entities are real because of their coherence and capacity to spark change. This is a strong argument for Chalmers' claim that virtual reality is a genuine reality, not a fiction. Whatever happens in the metaverse is important; living in it is purposeful.

How Is It Accessible?

As the book near its end, we may want to have a little throwback to the beginning of it. The metaverse is a multidimensional virtual environment that can be accessed through portal digital devices, smartphones, laptops, desktops, head-mounted displays, and games consoles. Certainly, the metaverse has also been accessed through other types of equipment (i.e., goggles, glasses, lenses, webcams, screens, and microphones). No matter what digital tool, it should be taken seriously, with or without the hype that surrounds it. The main significance of this is unmistakable: the metaverse is the epitome of the astronomical immersive shift in the datafication and informatization of daily human activities (for both business and pleasure). The mere possibility to observe a room full of interactants suggests that the metaverse can be accessed for beginners—in this case, as a visually-driven environment that does not offer promises for privacy or protection from data stealers. Of course, the metaverse has many more possibilities than the visual aspect of it. Although virtual realist governance underscores the lack of separation between the physical and the virtual, it has corollaries for the types of ethical issues that should be addressed with respect to phenomena such as the profiling of users in the metaverse (McStay, 2023b).

Since it has been demonstrated that using metaverse services is becoming easier and easier, it represents an incentive for laypeople to engage with it. In spite of the ocular bias of the metaverse, it is nevertheless a platform involving real participating bodies or those using photorealistic avatars. This biometric dimension of the metaverse is highly advanced, including facial recognition capabilities, as well as tracking precision for emotions, haptics, bodily movements, and brain patterns. For example, Meta, Microsoft, Apple, Amazon, and Roblox have developed superhuman-like skills in emotion recognition. This is one reason why Meta has invested a colossal amount of dollars in brain–computer interfaces via its Research Labs initiative. Now, a grave concern emerges when surveillance capitalism is the logical outcome. In many cases, when information collected about users' biometrics, neural activity, eye gaze, avatar behavior, and other past actions is added to a current database, one might apply the principles of virtual realist governance and reason that laws in such digital platforms should be imposed as those in the real world; that is, more regulation is needed to curtail surveillance of human behavior and interactions (McStay, 2023b).

Not Just Against Surveillance

Virtual realist governance is a theoretical concept that is not merely against surveillance, but it is also against design applications that motivate parties into taking actions that they would not otherwise engage in. From prior examples, we can reminder readers about problems such as unwarranted access to personal data, the resultant need for content moderation, inappropriate avatars and deepfakes, deceptive use of chatbots or advertising, and other strategies that aim at misleading users (McStay, 2023b). As Chalmers (2022) admits, the issue of the virtual is more often than not political. It deals not only with matters of democracy, anarchy, and corporate rule, but it also deals about the mechanics of how governance is implemented, especially when the digital is seen as unavoidably linked to long-established concerns of data protection. The key argument here is that this authentic surveillance of users is distressing from the standpoint of data protection and mental integrity. The authors of this work are particularly concerned about the subject-centered abuses of privacy, cognitive freedom, free agency, and first-person perspective (Lavazza, 2018; McStay, 2023a).

Hate Speech

The metaverse designs and promotes ways for users to connect virtually, allowing growth and profits to flourish, in conjunction with false information, surveillance, and a public occurrence that does not seem to be sufficiently highlighted as threatening: hate speech. As cited in Henson (2023), hate speech refers to "any kind of communication in speech, writing or behavior, that attacks or uses pejorative or discriminatory language with reference to a person or a group on the basis of who they are" (pp. 119–20). As he continues, it is "based on their religion, ethnicity, nationality, race, color, descent, gender, or other identity factor" (pp. 119–20). Some may assume that the pursuit of digital innovation should not take online verbal violence into account when it is viewed as a phenomenon with no effect on the real world. However, tragedies such as mass shootings and highly mediatized suicides show how unrestricted digital spaces can provide cannon fodder for those who promote—through words or manifestos—deadly violence. Technological progress in VR boosts the immersion of potentially toxic communicators by giving them more latitude in expressing themselves and interacting with like-minded others (Shariff et al., 2023).

Minority Groups as Targets

Most of the targets of digital hate speech are minority groups, particularly minority women, who are disproportionately victimized (Henson, 2023). Hate speech is nefarious due to its potential to normalize intolerance, which can put translate into grave outcomes. When hate speech is made to mushroom, it gives rise to the creation of norms: hatred and intolerance become second nature in society (Matusitz, 2012; McCardy & Matusitz, 2021). Under these circumstances, behaviors that are deemed extreme become conceivable and plausible (Futtner & Brusco, 2021). Another manner in which hate speech proliferates happens through algorithms, which excel at locating targets such as minority groups. The danger of algorithms is particularly pervasive not only because the algorithmic capabilities seem limitless but also because their ubiquity is what motivates extremists to exploit the platform and employ its recommendation system to steer audiences toward their ideologies (Henson, 2023).

As the metaverse is designed to virtually represent the physical world, attacking minority groups is expected to happen in many different ways. One such method is through hate speech and harmful language. Consequently, it is imperative to locate, acknowledge, and monitor these words (Medina et al., 2022). The challenge is that governments have varying societal values, concerning either freedom of expression (over censorship) or freedom to access any types of content (including harmful ones). States will apply their own laws and offer different levels of protection accordingly. In the absence of an overarching international apparatus, such as the

United Nations, it will remain difficult to adequately address hate speech and censorship of it, particularly since the latter is said, by some, to endanger the full concretization of freedom of expression (cited in Henson, 2023).

Mobbing

Ultramodern machines and deep learning algorithms, such as "natural language" applications, are being misused by mobs in the metaverse to spew deep-rooted hatred toward individuals or groups with opposing viewpoints. These algorithms exert influence on users through intermittent, ephemeral, and adapted experiences in ways that abolish opposing viewpoints. State-of-the-art immersive capabilities facilitate the luring of publics into believing what fits their preferences and make other "things" uncomfortable and rejectable. This, in turn, creates predicaments as certain groups of society (mostly youths) become less able to understand or respect the sentiments and attitudes of others, a poor social skill that could still be remedied through real face-to-face social interactions. This ideological predicament is akin to an echo chamber because users are conditioned into growing in a collective entity of identity construction and idea formation. Thus, this presents another challenge to academics who seek to develop fact-based recommendations or countermeasures to hate speech (Bibri & Allam, 2022).

These cognitive echo chambers are not only eased and promoted by the metaverse, but they also bear profound consequences on our own behavioral decisions regarding health and safety matters. For example, during the COVID-19 pandemic, a high proportion of individuals took draconian decisions because of the "absurd" mandates imposed on them by the states. In this case, conspiracy theories are a major conveyor of hate speech. Ethical principles that used to assist people in making sound everyday decisions have been significantly undermined due to the rise of digital platforms. A myriad faith-based, social, ideological, and political beliefs are inspired by the widespread use of digital platforms in which ideological frames operate as force multipliers (Bibri & Allam, 2022). These frames are based on the selection of certain aspects of a perceived reality, become more salient over time, form an agenda or grand narrative, encourage causal interpretation, and eventually shape people's moral evaluation (Matusitz, 2022). Such frames become social representations that are culture-specific and institutionalized within society. Once individuals are attuned to specific frames, they develop their values accordingly. Put simply, frames shape the way people think and what they should think about (Moscovici, 1984).

Keen (2016) writes against the dangers of mob mentality that creates frames in order to push their agendas on digital platforms. It empowers them rather than democratic processes and, thus, freedoms. The metaverse can easily aggravate the situation, as the world has added a great volume of immersive media to our current use of social media. The algorithms are generally opaque and invisible, leading to massive circles of influence in the metaverse. It is fair to say that big-tech corporations, and even traditional corporate media, allow the massive bulwark of hateful content through some form of mass formation psychosis—remember the terms "opaque" and "invisible" a few sentences earlier (making users subjects who can be effortlessly manipulated). These conditions impact the human mind in a manner that causes users to lose touch with reality and, therefore, become more oblivious to what is real. In due course, these people may even jump on the hate speech bandwagon by exhibiting disconcerting thoughts and distorted perceptions of the world. This is how mob influence works, and it has proven to be efficient in the metaverse.

Taken as a whole, the metaverse is bound to pose to challenges to critical thinking—wittingly or unwittingly. Upon looking at more details, this process may occur by suppressing or banning news, posts, debates, forums, videos, films, and other communicative artifacts that are regarded as politically objectionable. In these conditions, the perpetrators will mobilize highly influential algorithms and recommender systems. Before the advent of virtual platforms, content moderation was already shaping social norms and, at the same time, the social fabric and cultural production of a community. The difference with AI-based techniques is that changing the opinions, views, and beliefs of audiences is much more efficient and much faster. This makes ideologues' and propagandists' endeavors to fit their virtual communities a task that is so easy and fervent—with just a few mouse clicks or apps away. As we have seen, a significant percentage of the targeted groups will eventually neglect their own beliefs and principles by adopting the mob's behaviors on virtual platforms. Such hive mentality can also be sparked by the unbridled censorship of challenging political rhetoric. This is tantamount to the totalitarian control of narratives pushed by the ruling élite, turning modern society into one that relies on dominant modes of communication and on the apparatus of the powers-that-be (Bibri, 2022).

Diversity and Inclusivity as Solutions

The anonymity and rapid evolution of the metaverse favor the development of hate speech and mobbing behavior. Too many users feel empowered to undertake discriminatory behavior without suffering consequences. This inevitably leads to a toxic environment for minority groups (and other groups, for that matter). To tackle these issues, both big-tech corporations and metaverse aficionados must propose solutions based on diversity and inclusivity. This can include, inter alia, the design and implementation of avatars that appear through a diverse set of physical appearances, providing safe spaces for all parties involved. This would be in addition to integrating systems to address hate speech and harassment (Abdusatarov, 2023). Furthermore, it is important to educate users on the values of respect and inclusivity in the metaverse. For example, metaverse administrators could incorporate training programs for users that teach them how to identify and address discriminatory behavior. On a broader level, this could consist of setting community guidelines that stress inclusivity and respect for all. The metaverse has immense potential for inclusivity and diversity. In doing so, we can create a virtual environment that reflects the values of a truly inclusive world (Simi & Matusitz, 2022).

In line with these solutions, the very notions of diversity and inclusivity in the metaverse should be commensurate with the current laws in place. Although the metaverse is sufficiently large to accommodate diversity of thoughts and participants, it must honor the laws that have been passed to regulate virtual activities and interactions. On the face of it, it may seem like a simple criterion to meet; however, the complexity of jurisdictions and existing regulations may make certain forms of diversity difficult to implement. In particular, the concentration or centralization of governance may pose a challenge, given the ways the metaverse constantly evolves. The application of "law and order" in virtual environments and a system of checks and balances for regulators (e.g., punishment and reward mechanisms) could represent a solution to this state of perplexity. A design that is both ethical and legal is required to accomplish this (Sethi, 2022).

Lastly, when crime or unethical behavior in the real world gets transposed into the virtual world, their potency for ramifications becomes the subject of legal debate. Even in the event of a "gray area," which is often the case in debates about convergence between the physical and the virtual, crime or unethical behavior in the metaverse needs to be evaluated on its own uniqueness. Comparing a violation committed online to one in person will remain a controversial matter. An illustration of this would be an incident in which a digital avatar gets raped after entering a digital game or similar online setting. Such incident would require a different jurisdiction to address the type of tort and the severity of punishment to be applied. Conversely, the perpetrators' defense would likely be that the metaverse does not have (1) a physical benchmark to make any evaluation of improper physical behavior (such as rape) and (2) enough diversity and inclusivity to let users say or do what they wish. More thoughts should be considered before serious conclusions to such problems can be legitimately drawn (Sethi, 2022).

References

Abdusatarov, J. (2023). Issues that need to be resolved when developing the legal framework of international private law relations in metaverse. *World Bulletin of Management and Law, 21*, 193–204.

Ahmed, F., Zubair Shafiq, M., & Liu, A. X. (2016). The internet is for porn: Measurement and analysis of online adult traffic. *IEEE International Conference on Distributed Computing Systems, 36*, 88–97. https://doi.org/10.1109/ICDCS.2016.81

Ali, M., Naeem, F., Kaddoum, G., & Hossain, E. (2022). *Metaverse communications, networking, security, and applications: Research issues, state-of-the-art, and future directions*. arXiv.org. Retrieved on May 24, 2023 from https://arxiv.org/pdf/2212.13993.pdf

Al-Shareeda, M. A., & Manickam, S. (2022.): Article 1543). Man-in-the-middle attacks in mobile ad hoc networks (MANETs): analysis and evaluation. *Symmetry, 14*(8). https://doi.org/10.3390/sym14081543

Baptista, E. (2019). China's communist party is making its own (virtual) reality. *Foreign Policy*. (June 21, 2019). Retrieved on May 19, 2023 from https://foreignpolicy.com/2019/06/21/chinas-communist-party-is-making-its-own-virtual-reality/

Berners-Lee, T. (2018). The web can be weaponised—And we can't count on big tech to stop it. *The Guardian*. (March 11, 2018): A1. Retrieved on October 25, 2022 from https://www.theguardian.com/commentisfree/2018/mar/12/tim-berners-lee-web-weapon-regulation-open-letter

Bibri, S. E. (2022). The social shaping of the metaverse as an alternative to the imaginaries of data-driven smart cities: A study in science, technology, and society. *Smart Cities, 5*(3), 832–874. https://doi.org/10.3390/smartcities5030043

Bibri, S. E., & Allam, Z. (2022). The metaverse as a virtual form of data-driven smart urbanism: On post-pandemic governance through the prism of the logic of surveillance capitalism. *Smart Cities, 5*(2), 715–727. https://doi.org/10.3390/smartcities5020037

Bojić, L. (2022a). *Culture organism or techno-feudalism: How growing addictions and artificial intelligence shape contemporary society*. Institute for Philosophy and Social Theory.

Bojić, L. (2022b.): Article 22). Metaverse through the prism of power and addiction: what will happen when the virtual world becomes more attractive than reality? *European Journal of Futures Research, 10*(1). https://doi.org/10.1186/s40309-022-00208-4

Cadwalladr, C., & Graham-Harrison, E. (2018). How Cambridge analytica turned Facebook 'likes' into a lucrative political tool. *The Guardian*, A1. Retrieved on October 25, 2022 from https://www.theguardian.com/technology/2018/mar/17/facebook-cambridge-analytica-kogan-data-algorithm

Chalmers, D. (2022). *Reality+: Virtual worlds and the problems of philosophy*. Allen Lane.

Couldry, N., & Mejias, U. A. (2019). Making data colonialism liveable: How might data's social order be regulated? *Internet Policy Review, 8*(2), 10–21. https://doi.org/10.14763/2019.2.1411

Crutzen, C. K. M. (2005). Intelligent ambience between heaven and hell: a salvation? *Journal of Information, Communication and Ethics in Society, 3*(4), 219–232. https://doi.org/10.1108/14779960580000275

Curtis, C., & Brolan, C. E. (2023). Health care in the metaverse. *The Medical Journal of Australia, 218*(1), 46. https://doi.org/10.5694/mja2.51793

Egliston, B., & Carter, M. (2021). Critical questions for Facebook's virtual reality: Data, power and the metaverse. *Internet Policy Review, 10*(4), 1–23. https://doi.org/10.14763/2021.4.1610

Falchuk, B., Loeb, S., & Neff, R. (2018). The social metaverse: Battle for privacy. *IEEE Technology and Society Magazine, 37*(2), 52–61. https://doi.org/10.1109/MTS.2018.2826060

Futtner, N., & Brusco, N. (2021). *Hate speech is on the rise*. Geneva International Centre for Justice.

Gray, J. (2019). The new tech totalitarianism. *The New Statesman*. (February 6, 2019). Retrieved on September 20, 2022 from https://www.newstatesman.com/culture/2019/02/the-new-tech-totalitarianism

Heim, M. R. (1998). *Virtual Realism*. Oxford University Press.

Henson, A. (2023). Virtual Whac-A-Mole: Addressing the patchwork regulation of online hate speech. *Michigan State International Law Review, 31*(1), 116–156. https://doi.org/10.17613/aky5-jz66

Hills-Duty, R. (2018). *China's Communist Party uses VR for loyalty test*. VR Focus. (May 15, 2018). Retrieved on May 19, 2023 from www.vrfocus.com/2018/05/chinas-communist-party-uses-vr-for-loyalty-tests/

Hine, E., & Floridi, L. (2022). New Deepfake regulations in China are a tool for social stability, but at what cost? *Nature Machine Intelligence, 4*(7), 608–610. https://doi.org/10.1038/s42256-022-00513-4

Hu, Y., & Liu, C. (2022). The 'Metaverse society': Beyond the discourse intrinsic potential and transformative impact. *Asia Pacific Academy of Science, 3*(2), 1–14. https://doi.org/10.54517/met.v3i2.2128

Huq, N., Reyes, R., Lin, P., & Swimmer, M. (2022). Metaverse or metaworse: cybersecurity threats against the internet of experiences. *Trend Micro*, 1–24. Retrieved on May 30, 2023 from http://

documents.trendmicro.com/assets/white_papers/wp-metaverse-or-metaworse-cybersecurity-threats-against-the-internet-of-experiences.pdf

Ivanova, E., & Borzunov, G. (2020). Optimization of machine learning algorithm of emotion recognition in terms of human facial expressions. *Procedia Computer Science, 169*, 244–248.

Kang, G., Koo, J., & Kim, Y.-G. (2023). Security and privacy requirements for the metaverse: A metaverse applications perspective. *IEEE Communications Magazine*, 1–7. https://doi.org/10.1109/MCOM.014.2200620

Keen, A. (2016). The Internet Is Not the Answer (New York: Grove Press).

Lavazza, A. (2018). Freedom of thought and mental integrity: The moral requirements for any neural prosthesis. *Frontiers in Neuroscience, 12*(82), 10–21. https://doi.org/10.3389/fnins.2018.00082

Li, X., He, J., Vijayakumar, P., Zhang, X., & Chang, V. (2022). A verifiable privacy-preserving machine learning prediction scheme for edge-enhanced HCPSs. *IEEE Transactions on Industrial Informatics, 8*(8), 5494–5503. https://doi.org/10.1109/TII.2021.3110808

Marwick, A. E., & Boyd, D. (2011). I tweet honestly, I tweet passionately: Twitter users, context collapse, and the imagined audience. *New Media & Society, 13*(1), 114–133. https://doi.org/10.1177/1461444810365313

Matusitz, J. (2012). Relationship between knowledge, stereotyping, and prejudice in interethnic communication. *PASOS: Journal of Tourism and Cultural Heritage, 10*(1), 89–98. https://doi.org/10.25145/j.pasos.2012.10.008

Matusitz, J. (2022). *From child terrorism to peace activism*. Springer.

McCardy, A., & Matusitz, J. (2021). Power in hidden figures: A critical discourse analysis. *International Journal of Media & Cultural Politics, 17*(1), 3–17. https://doi.org/10.1386/macp_00035_1

McStay, A. (2023a). *Automating empathy: When technologies are claimed to feel into everyday life*. Oxford University Press.

McStay, A. (2023b.): Article 13). The metaverse: Surveillant physics, virtual realist governance, and the missing commons. *Philosophy & Technology, 36*. https://doi.org/10.1007/s13347-023-00613-y

Medina, R. M., Njoku, J. N., Lee, J.-M., & Kim, D.-S. (2022). Audio-based hate speech detection for the metaverse ising CNN. *Korean Institute of Communications and Information Sciences*. Retrieved on June 11, 2023 from https://www.researchgate.net/publication/365877973_Audio-Based_Hate_Speech_Detection_for_the_Metaverse_using_CNN

Miller, M. R., Herrera, F., Jun, H., Landay, J. A., & Bailenson, J. N. (2020.): Article 17404). Personal identifiability of user tracking data during observation of 360-degree VR video. *Scientific Reports, 10*(1). https://doi.org/10.1038/s41598-020-74486-y

Moscovici, S. (1984). The phenomenon of social representations. In R. Farr & S. Moscovici (Eds.), *Social representations* (pp. 3–69). Cambridge University Press.

Needleman, S. E. (2021). The amazing things you'll do in the 'metaverse' and what it will take to get there. *The Wall Street Journal*, A1. Retrieved on August 24, 2022 from https://www.wsj.com/articles/the-amazing-things-youll-do-in-the-metaverse-and-what-it-will-take-to-get-there-11634396401

Ometov, A., Bezzateev, S. V., Kannisto, J., Harju, J., Andreev, S., & Koucheryavy, Y. (2017). Facilitating the delegation of use for private devices in the era of the internet of wearable things. *IEEE Internet of Things Journal, 4*(4), 843–854. https://doi.org/10.1109/JIOT.2016.2593898

Parasol, M. (2022). China's digital yuan: Reining in Alipay and WeChat pay. *Banking & Finance Law Review, 37*(2), 265–303.

Pooyandeh, M., Han, K.-J., & Sohn, I. (2022.): Article 12993). Cybersecurity in the AI-based metaverse: A survey. *Applied Sciences, 12*(24). https://doi.org/10.3390/app122412993

Rosén, J., Kastrati, G., Reppling, A., Bergkvist, K., & Åhs, F. (2019). The effect of immersive virtual reality on proximal and conditioned threat. *Scientific Reports, 9*(1). https://doi.org/10.1038/s41598-019-53971-z

Rosenberg, L. B. (2021). Augmented reality: Reflections at thirty years. In *Proceedings of the Future Technologies Conference (FTC)* (Vol. 1, pp. 1–11). Kohei Arai. Springer.

Rosenberg, L. B. (2022). Regulation of the metaverse: A roadmap. *International Conference on Virtual and Augmented Reality Simulations (ICVARS), 6*, 1–11. https://doi.org/10.1145/3546607.3546611

Ruiz Mejia, J. M., & Rawat, D. B. (2022.): Article 21882815). Recent advances in a medical domain metaverse: Status, challenges, and perspective. *International Conference on Ubiquitous and Future Networks (ICUFN), 13*. https://doi.org/10.1109/ICUFN55119.2022.9829645

Salau, A., Dantu, R., Morozov, K., Upadhyay, K., & Badruddoja, S. (2022). Towards a threat model and security analysis for data cooperatives. *International Conference on Security and Cryptography, 19*, 707–713. https://doi.org/10.5220/0011328700003283

Sethi, A. (2022). Security and privacy in metaverse: Issues, challenges, and future opportunities. *Cyber Security Insights Magazine, 2*, 1–4.

Shang, J., Chen, S., Jie, W., & Yin, S. (2022). ARSpy: Breaking location-based multi-player augmented reality application for user location tracking. *IEEE Transactions on Mobile Computing, 21*(2), 433–447. https://doi.org/10.1109/TMC.2020.3007740

Shariff, S., Dietzel, C., Macaulay, K., & Sanabria, S. (2023). Misogyny in the metaverse: Leveraging policy and education to address technology-facilitated violence. In H. Cowie & C.-A. Myers (Eds.), *Cyberbullying and Online Harms*. Routledge. Chapter 10.

Simi, D., & Matusitz, J. (2022). Improving cultural diversity in U.S. higher education: General perspectives. *International Journal of Social Sciences & Educational Studies, 9*(4), 117–129. https://doi.org/10.23918/ijsses.v9i4p117

Venturini, T., & Rogers, R. (2019). 'API-based research' or how can digital sociology and journalism studies learn from the Facebook and Cambridge analytica data breach. *Digital Journalism, 7*(4), 532–540. https://doi.org/10.1080/21670811.2019.1591927

Wasserkrug, S., Gal, A., & Etzion, O. (2008). Inference of security hazards from event composition based on incomplete or uncertain information. *IEEE Transactions on Knowledge and Data Engineering, 20*(8), 1111–1114. https://doi.org/10.1109/TKDE.2008.74

Waters, A. (2021). Will neoliberal capitalism survive the coronavirus crash or is this the beginning of techno-feudalism? *Journal of Australian Political Economy, 86*, 406–431.

Wei, J., Li, J., Lin, Y., & Zhang, J. (2021). LDP-based social content protection for trending topic recommendation. *IEEE Internet of Things Journal, 8*(6), 4353–4372. https://doi.org/10.1109/JIOT.2020.3026366

Wei, C., Lin, W., Liang, S., Chen, M., Zheng, Y., Liao, X., & Chen, Z. (2022.): Article 131). An all-in-one multifunctional touch sensor with carbon-based gradient resistance elements. *Nano-Micro Letters, 14*. https://doi.org/10.1007/s40820-022-00875-9

Wiggers, K. (2022). How the Metaverse could transform upskilling in the enterprise. *Venture Beat*. (January 26, 2022). Retrieved on February 9, 2023 from https://venturebeat.com/ai/how-the-metaverse-could-transform-upskilling-in-the-enterprise/

Williams, G. A. (2023). As global hype wanes, can China pull ahead in the metaverse race? *Vogue Business*. (May 19, 2023). Retrieved on June 11, 2023 from https://www.voguebusiness.com/technology/as-global-hype-wanes-can-china-pull-ahead-in-the-metaverse-race

Yao, P., Lympouridis, V., & Zyda, M. (2021.): Article 20592810). Virtual equipment system: Face mask and Voodoo doll for user privacy and self-expression options in virtual reality. *IEEE Conference on Virtual Reality and 3D User Interfaces Abstracts and Workshops*. https://doi.org/10.1109/VRW52623.2021.00256

Zuboff, S. (2015). Big other: Surveillance capitalism and the prospects of an information civilization. *Journal of Information Technology, 30*(1), 75–89. https://doi.org/10.1057/jit.2015.5

Chapter 10
Discussion and Future Directions

What this book has demonstrated is that communication in the metaverse is proof that the world is increasingly transitioning to the digital era. The metaverse opens the door for multisensory interactions not only among people but also among digital ecosystems, physical entities, and digital twins. The universe is where time, space, and matter interact, where all things exist, and where humans play only a small part in it. The metaverse can easily evolve from a single natural world to a dual world of virtual and real coexistence. As the twin of the real world, the metaverse exploits and integrates the evolutionary and representational character of the physical world, and altogether, thrives on the immensely influential and heavily scientific and technological ability to concretize the landing of creativity and imagination.

It is the next generation where our social presence and our avatars can enhance our immersion in virtual environments with the help of human–computer interaction by mirroring an increasingly broader range of motions and emotions. The relationship between people and digital information is constantly evolving. On the one hand, people can acquire an enormous amount of information from texts, images, and videos in various ways. On the other hand, from head-mounted displays, touch screen, and virtual voice, the interactive modes of information are getting closer to our five natural senses. The popularity of the metaverse will constantly promote the evolution of social and interpersonal relationships. Today's forms of communication will move toward a quasi-total integration of digital and real communication (Liu & Zheng, 2022).

How to effortlessly create a personalized digital human, how to mimic a real human's movements and emotions to a virtual human, how to construct identities, how to cooperate, and how to educate have become major areas of research in academia, private sector, and public institutions (Zhang et al., 2022). The main conclusion of this book is that communication in the metaverse rests on four principles. As such, it is (1) a complex interactive environment, (2) a socially shaped environment, (3) a controlled media environment, and (4) a disruptive innovation. The latter

J. Matusitz, J. Dacas, *Communication in the Metaverse*, Signals and Communication Technology, https://doi.org/10.1007/978-3-031-63485-7_9

principle, disruptive innovation, can be broken down into four categories: disruptive communication, a higher capacity for deception, social engineering, and privacy concerns and surveillance. The second half of this chapter provides detailed suggestions for future research.

A Complex Interactive Environment

In the metaverse, psychological distance is not commensurate with geometric distance. It is important to regard our experience of space as cocreated meaning. More precisely, being in a space is a matter of empathic interaction, where the space and its users cocreate mood. In virtual environments, this process is holistic and produced through the copresence of individuals with their own interpretations of space; this justifies, in part, why the metaverse resists a singular or immutable description (cited in Paananen et al., 2022). In fact, the metaverse is so complex that, unlike communication in the three-dimensional world, where both oral and written communication slowly evolve decade by decade, the innovations and evolutions of digital platforms happen so fast that they motivate human beings to push themselves to increasingly higher summits, breaking through the shackles of the real world, and exposing their bodies, minds, and identities to unprecedented possibilities for communication and social interaction (cited in Wang & Wei, 2022). This is particularly evident in healthcare contexts, where health workers can improve patients' conditions through various means of virtual collaborations.

Outside professional contexts, at the simpler level, the metaverse greatly facilitates communication between interactants. In traditional face-to-face communication, human beings rely on natural elements such as bodily movements and facial expressions to spread information in a small range. The invention of characters quickly allowed human beings to expand the dimension of geography and body; now, they could rely on paper and other carriers for long-distance communication. The invention of broadcasting was a milestone in the history of humankind because it freed us from the burden of both hands and feet, and pushed us to new heights with the planetary dissemination of information (even though broadcasting was emitted only through oral discourse). Today, we operate in a digital world built through digital symbols and images. These symbols and images are replacing (or at least improving) our ways of disseminating information. Our body has been liberated in manners never seen before; and the distinction between communicative activities in that universe and physical body have been greatly reduced (cited in Wang & Wei, 2022). As Katherine Hayles said: "The represented body appears on one side of the computer screen as flesh and blood, and the reproduced body is generated in the electronic environment through language and semiotic marks" (cited in Wang & Wei, 2022, p. 7).

A Socially Shaped Environment

Even though the metaverse is not a biological space in which we have evolved, it should be acknowledged that it is evolving around us in a universe parallel to the biological norm. This increases the potential to produce a person-world akin to our three-dimensional world. It is a dimension that meets the needs and instincts of the individuals within it. This new reality of the social and physical allows our actions, interpretations, beliefs, and habits to forge new communities, institutions, and hierarchies. The interactions among users, as well as those between users and objects in virtual environments, strengthen or build on such communities, institutions, and hierarchies—year after year. By extension, such as in the physical space, the relationship between the early metaverse adopters and the subsequent (late) bloomers hinges, in part, on transferring rituals and customs—best tools, tips, or skills on how to properly communicate in the metaverse—during which they gradually become better (Davidson, 2022).

A theory that illustrates the above is social shaping of technology (SST). SST is a theoretical model that seeks to describe technological change as a social process. It is predicated on the notion that technology and society are mutually inclusive, rather than separate entities, and that each shapes the other in highly correlated and multifaceted ways (in contrast to linear pathways). From the vantage point of SST, social and technical are inseparable (MacKenzie & Wajcman, 1999), and each evolves in tandem with the other. Technologies, then, are social products, and every phase of evolution of new technologies depends on a scope of social factors that affect the nature of the technologies and their corollaries (Williams & Edge, 1996). The SST approach challenges technological determinism. The term technological determinism is "an approach that identifies technology or technological developments, as the central causal element in processes of social change" (Croteau & Hoynes, 2019). This implies a type of social change so profound that it will regulate not only the media environment but also the entirety of our lives. Put another way, the premise is that, unless we adjust to a technological system, we cannot function as a society (Mansell, 2012).

SST rejects the simple causality of technological determinism, while recognizing that it is partially correct in its notion that technology matters. The idea of the metaverse as a socially shaped environment embodies the reality that technologies influence change over time, but humans—or at least the powers-that-be (see next section)—still have agency over the methods by which they are developed (Mansell, 2012). The clever selection of the term "shaping," as opposed to "social construction," highlights the materiality of technology (Vasist & Krishnan, 2023). SST theorists argue that the interactions between social forces and the physicality of technologies merit serious consideration (Howcroft & Taylor, 2023). From the SST perspective, technologies mirror particular forms of power and authority and impact how we perform our daily tasks, communicate, and consume over time (Winner, 1980).

A Controlled Media Environment

The powers-that-be controlling the media environment have immense power to shape, enlighten, and influence audiences. The metaverse operates on similar principles. For people embedded within a full sensory experience, the lessons learned (and the experiences lived) have the potential to penetrate deeply into the participants' identity and psyche (Murray, 2022). As readers know by now, the metaverse is a product of both private companies and established systems, shaped by society. In this context, communication, significance, information, and our comprehension of science, technology, and our environment all rely on social influences. As a form of technological activity, it is also controlled and regulated by sociopolitical and economic structures (Bibri, 2022). If truth be told, the metaverse could offer various social and information environments with the same level of freedom, opportunity, and interactivity commonly found in video games (Nalbant & Aydin, 2023).

However, it is still owned or overwhelmingly regulated by diverse actors, politicians, and CEOs, thereby affecting the degree of freedom within such social and information ecosystems. As we have seen in Chap. 3, when describing intercultural communication between metaverse inhabitants, the fact that meanings within the metaverse—or even meanings of what the medium should be—are not always easy to grasp can lead to misunderstandings, misconceptions, and illogical viewpoints. Key actors may be inclined to put their stamp on meanings of the metaverse or how it should be monopolized, thereby attempting to control the medium for their own developments. We do not believe that an ultimate or agreed-upon definition of the metaverse will ever emerge in the public discourse (Dolata & Schwabe, 2023). This is why we need to understand which aspects and visions are associated with the metaverse across the globe and which actors, politicians, and CEOs seek to shape it.

A Disruptive Innovation

Disruptive innovations are not necessarily against the idea of improving technologies. In fact, history teaches us that, in the early twentieth century, the automobile was a disruptive innovation to the horse-drawn carriage (Hardman et al., 2013). Instead, they disrupt and redraw the current market trajectory by including both opportunities (and/or products and services) and threats (Horn & Staker, 2014). Forecasts can be made on how imminent certain disruptive technologies will concretize and what unique features they may have, depending on all sorts of societal considerations. These factors are not necessarily technological; they are often external factors that fall outside the scope of technology forecasting (Bibri, 2018).

Disruptive Communication

The metaverse disrupts because it does not merely create new meaning, but it also changes the rules of human meaning. History repeats itself, particularly when language replaced rudimentary biological communication, and when writing improved speech, all of which demonstrating constant cultural evolution. Nowadays, technological advancement is chipping away at this margin of safety of human primacy, not only because AI recognizes faces with a rapidity and confidence unparalleled by humans—in situations with low visibility, with a mask on, when moving our bodies, etc.—but also and above all because AI recognition capabilities are acquiring an ever-increasing human capacity for communication and cognition (Leone, 2023).

This is a throwback to the discussions about avatars, impression management, and identity creation. There is no area within the metaverse that more captivates and, at the same time, distresses humans than the one where the capacity of AI exceeds the average person's ability of imitation. Until the development of AI, machines in the past were more cognitively powerful than we were, but fundamentally incapable of imitating us with respect to our most unique traits. Animals such as African grays and apes, on the other hand, can imitate some traits of human behavior very closely, but with a cognitive ability that is strikingly inferior and, on most occasions, not very versatile. Today, avatars and other computing models of communication are causing inquietude because they not only exceed us in technological feats, but they also begin to imitate us at alarming levels (Leone, 2023).

Other concerns have been raised. For instance, what happens to haptics (i.e., touching) if the connection between virtual interaction and physical erogenous regions becomes flexible or adaptable? In a similar fashion, the handshake may excite more erogenous regions of our body via remapping (Davidson, 2022). The issue of interpersonal trust becomes supreme. Digital sexual intercourse and romantic relationships would necessitate more secure, password-protected spaces. For a user to see or interact with other users would require getting the permission codes of his or her profile. As Rheingold (1991) put it, "it might be that the physical commingling of genital sensations will come to be regarded as a less intimate act than the sharing of the data structures of your innermost self-representations" (p. 352).

As some critics would argue, irrespective of how advanced communication becomes in the metaverse, there should be a clear distance and separation between the digital and the physical. If this is not the case, what may happen is that the receiver could interpret the virtual sign as a physical entity; then, it will stop being a virtual sign and, instead, carry a meaning different from what was anticipated. Such a situation will give rise to incongruence and disparity within subsequent communications. For example, unaware that humans can fly in VR but not in the real world, a user who is so entrenched within the metaverse may misunderstand this and could end up dying. This is evocative of people who are fanatics of celebrities and believe their on-screen personas are actually real. The logical communication breaks down because such fanatics cannot distinguish between different worlds (Davidson, 2022).

A Higher Capacity for Deception

As discussed in detail in Chap. 8, the metaverse is also disruptive in its ability to magnify the possibilities for deceiving users. Disinformation was a keyword of that chapter. Disinformation can happen through targeted disinformation campaigns meant to bemuse society for political gains, making deliberately flawed information appear as valid, particularly since it has been ascertained that bots excel at operating repetitive tasks such as spouting automated messages, applying user following techniques, and creating fake accounts that lure users to dangerous territories. Unfortunately, false information can spread six times as fast as authentic content. This explains how human security attacks are so easily churned out (Repez & Popescu, 2020).

Disinformation raises concerns regarding polarization, dwindling trust, and erosion of democratic values. With recent technological progress in VR and AI, academics have focused on how to tackle these issues and how to prevent any potential damage they have on current and future generations (Brown et al., 2023). In line with these contentions, deception in the metaverse may badly affect users' ability to recollect events, thereby leading to misperceptions. For example, Loftus's (2005) misinformation effect refers to "the impairment in memory for the past that arises after exposure to misleading information" (p. 361). As a specific example, a person who is targeted by false information can have a more difficult time remembering a prior experience with accuracy. Considering that virtual environments were initially designed to create unique experiences, they have the unfortunate potential to transform memories of real-world experiences, contributing to misperceptions (Brown et al., 2023).

Social Engineering

Humans make their own choices to engage with the metaverse voluntarily; there is no implication of coercion. However, this autonomy does not grant immunity from falling victim to social manipulation. The chief reason behind the remarkable effectiveness of social engineering lies in its subtle "soft power" approach to communication, enticing individuals into addictive behaviors related to specific metaverse activities or applications. The recommendation algorithm plays a pivotal role in this, effortlessly enticing metaverse users. Hence, regulating the metaverse becomes crucial for the global future. This advanced technology operates with sophistication, crafting tailored suggestions by manipulating user data. The prevalence of fake news and other manipulative content has had a perilous impact on societies worldwide, as these harmful materials are promoted by recommender algorithms, stirring negative emotions (Bojić, 2022b).

Promoting content that triggers discussions, amplifies emotional responses, and keeps people engaged on social media platforms typically involves the promotion of conspiracy theories and populist material. This, in turn, fosters the formation of echo chambers, social disconnection, heightened anger, and the emergence of conflicts within society. People tend to seek validation for their beliefs from those who share their viewpoints, leading them to encounter even more content that aligns with their existing social media consumption. This pattern further radicalizes specific groups, fosters hate speech, and deepens the divisions within societies, ultimately creating isolated echo chambers. This straightforward process is a significant driver of polarization, a prevailing issue in today's global societies (Bojić, 2022a).

Privacy Concerns and Surveillance

Chapter 9 dealt with additional problems, including privacy concerns and surveillance. The diversity and abundance of activities available in the metaverse are appealing to users of all ages, particularly youths between 13 and 18 years old. Their presence across multiple metaverse platforms is often seen through their rather shorter avatars (Blackwell et al., 2019) and/or their higher-pitched voices (Maloney et al., 2020a, b). For these youths, the metaverse offers phenomenal opportunities for social participation, enjoyment, and interactive play. At the same time, it raises grave concerns regarding their well-being and safety. Considering that the metaverse enables teenagers and young adults to develop intimacy and emotional connections, encounter deep interpersonal interactivity beyond gameplay, and engage in nuanced group activities, ill-intentioned users and would-be "prowlers" can exploit these preys by getting them to share sensitive information, expose them to mature content, and make them targets of virtual sexual assault (Maloney et al., 2020a, b).

These circumstances elicit the necessity of devising interactive, safe, and emotionally fulfilling metaverse platforms for users while alleviating the potential dangers of being preyed on, prowled, or surveilled. With the advancement of technologies, the proportion of people who use the metaverse increases and, as actions on a comparable scale to reality are executed, a great many bits of information are produced. Because metaverse data is more often than not valuable, and because its sheer quantity grows exponentially, the importance of security and dependability should not be called into question. When AI offers more guarantee to uphold diversity and respect, only then can it be considered an exemplary mechanism for allowing humans to connect in the metaverse (Maloney, 2021).

Directions for Future Research

In spite of the myriad capabilities, applications, and services afforded by the metaverse, it is still not sufficiently predictable, irrespective of the time horizon established for the dream of a global metaverse industry of US$ 1.3 trillion by 2030 (see https://www.precedenceresearch.com/metaverse-market). This figure gives us enough latitude to tackle interactive and safety concerns and address the difficult question as to how they will be made to dwindle. Prior studies have demonstrated that neglecting the disciplines related to technological forecasting explains why forecasts go wrong (Bibri & Krogstie, 2020). To draw a parallel, a major explanation for the ease with which Al-Qaeda successfully carried out the September 11, 2001 terrorist attacks was the US authorities' lack of forecasting or, to use a better a term, a lack of imagination. As Bougen and O'Malley (2009) explain, "official assessment of the circumstances surrounding 9/11 highlighted a lack of imagination in security policy as the major contributing factor to what had transpired" (p. 101).

Whether the problem pertains to a lack of technological forecasting or a lack of imagination, there ought to be more mechanisms to avoid any catastrophic failure in the domains of interaction and safety in the metaverse. In tandem with such efforts, there should be simpler or more transparent processes for increasing public participation in the influence or management of certain applications within metaverse, not merely control of our own profiles or avatars. This is particularly crucial when participants can bring effective insights, inputs, or informed choices on such virtual platforms. This process of public participation could even be done before any platform becomes mainstream because of the uses and abuses of big-tech corporations; they control various data-driven applications, social media, and recommender systems that impact a few billion people worldwide (Egliston & Carter, 2021).

What does this mean for communication practitioners and scholars? They should endeavor to be dexterous in both their technological and communicative approaches, be open to adopting new ones, and be willing to take risks in their efforts to forecast, let alone comprehend, the metaverse. Frohlich (2022) puts it elegantly when he explains that specialists should not look at the metaverse as a space that seeks to mirror reality but as an "activation channel." That is, it is an interactive environment that allows communication through avatars; it is an integrated communications strategy that facilitates the emergence of new personas and new data. Conversely, it also facilitates new types of problems, such as social engineering, surveilling, and bullying. The notion of activation channel is that of a new ecosystem, one that offers unprecedented opportunities to engage in two-way symmetrical communication or communicate in new contexts. Thus, to answer the question raised at the beginning of this paragraph, practitioners and scholars are in an ideal position to shape communication approaches and regulations within the metaverse, while bringing their research agendas to new heights.

Communication in the Workplace

Chapter 6 provided a detailed account on communication, social interaction, and collaboration in the workplace. Now, a few important questions loom on the horizon: Is the condition of an employee in the metaverse that of one who works at a virtual desk, using a virtual keyboard and screen to fulfill virtual duties? By extension, who decides how workplace collaboration is to be done in the metaverse? Addressing these questions is essential because, given the wide diversity of workers (Simi & Matusitz, 2022), it would be logical to deduce that they do not share the same motivation or aptitude to go "full metaverse." Considering the potential labor crowdsourcing problem in nations where labor costs are insignificant, organizations may resort to "social dumping" of their own workers (in their home states) in order to attract offshore employees who only need fast and effective training—bestowing them with reasonable communication skills to operate in virtual environments. To boot, ethical concerns also arise regarding the system of regulations and the algorithms that should be carefully selected to administer the labor market. This is particularly relevant for a situation in which human and artificial intelligence will have to interface and, in the process, influence each other (with both positive and negative consequences) (Caragnano, 2023).

The metaverse engenders a whole new set of working relationships developed through virtual platforms (with avatars, etc.). No matter how close they look to the real world, they nevertheless—to this day—remain dissimilar to our familiar, traditional normative patterns of communication and behavior. Contractual institutions should always keep this in mind. The same can be argued about new work patterns, including digitally driven forms of self-employment. Metaverse work patterns are becoming increasingly popular across the world, even though such nontraditional or nonstandard work configurations are not necessarily accepted or entirely subsumed under certain régimes. Even in Europe and the Western hemisphere, metaverse work patterns do not always align with traditional forms of employment. This is especially the case for full-time, permanent service or labor within the structure of a salaried and bilateral work relationship. Given these facts, more research needs to be conducted on the notion of digital workers because they will still be more and more immersed in a decentralized virtual workspace. In this situation, digital workers are like nomads, constantly moving and geographically disconnected from similar others (or even from the organizations that pay their salaries) (Caragnano, 2023).

Tackling the Digital Divide

Digital divide is the unequal access to computers, digital connections, and/or training in how to understand and apply these technologies. Second, it tackles the gap between the haves (people with regular, efficacious access to digital and information technologies) and the have-nots (people without this access). Third, it addresses the skills necessary to reach an adequate level of digital literacy. The metaverse cannot be used by a significant percentage of disadvantaged and isolated groups due to the digital divide. Consequently, these technological barriers have maintained their digital literacy at a lower level, thereby impacting their lives (Lythreatis et al., 2022). The global digital divide generates substantial differences in opportunity to access key virtual platforms in the domains of healthcare, education, and business. These disparities in access are particularly salient between developed and developing regions (Matusitz & Musambira, 2013; Musambira & Matusitz, 2015). The digital has been lionized as a great equalizer, a groundbreaking technological tool for the efficient sharing of information on a global level. However, geographical, sociopolitical, and ideological hurdles are still pervasive today, complicating the vision of digital gurus to have their inventions reach out as many people as possible. At a simpler level, one major obstacle is the difficulty to help disadvantaged and isolated groups acquire computers as well as corresponding networking and communications technologies (Ragnedda, 2017).

Future research should concentrate on examining the impact of helping developing regions access the metaverse and acquiring digital literacy skills. Currently, the self-referential approach to digital literacy—where one can learn these skills within the space itself or within the very community that needs it—could be insufficient in comparison with the educational models used in developed regions. The need for additional research into a metalanguage or metaverse for the acquiring of new forms of digital literacy would be more adequate (Davidson, 2022). Hence, the authors of this work advocate e-inclusion via access to the metaverse. Such digital inclusion would be made up of rules and guidelines that integrate activities leading to a more inclusive information society. In this fashion, new advancements in technology turn the difficulties associated with digital complexity into digital cohesion and openness. This ushers the presence of all things digital (i.e., platforms and related technologies) into all population strata, including underprivileged segments. Such digital cohesion and openness would also bear fruits in the realms of education (i.e., a subcategory called e-competence), age (i.e., e-ageing), disabilities (i.e., e-accessibility), gender, race, and minority statuses associated with remote regions (Matusitz, 2022).

From a financial standpoint, the reducing costs of key metaverse tools, such as head-mounted virtual-reality devices and technological products (e.g., Horizon Home, Future of Work, Presence Platform) will enable a considerably high number of people to immerse themselves in numerous virtual environments. Thus, the cost aspect may no longer be seen as an impediment to the wide acceptance of the metaverse. Statistics show the amount of money that users need to spend on state-of-the-art VR gear that is necessary to physically feel what can be experienced in the metaverse. It is increasingly affordable (https://www.statista.com/topics/8652/metaverse/). It is highly likely that certain segments of society—given their levels of education, socioeconomic status, and age—will be locked out (Bibri, 2022).

Of course, even a few decades from now, it is expected that the digital divide will continue to exist to some degree; the metaverse will still engender digital gaps and favoritism. This discussion on accessibility stratification capacity or discriminatory access in technology use will be perpetual; it will not remain ephemeral. Part of the reason is that tech-savvy leaders are the ones who will continue to set the course for technology adoption. This reality does not lend itself to full access for all. This even includes the non-technologically inclined citizens or the pre-laptop generations in developed regions (or other people who are hindered for diverse reasons). This will continue to be the case with the use of any novel technology, despite the promise made by the Meta company that developing youth's digital literacy programs could become a panacea (Bibri, 2022).

Final Thoughts

The fulfillment of the metaverse vision of a bright future for all remains elusive, and a significant portion of it will develop in unpredictable ways. Socio-technical leaders pledge for a better world, sometimes with a clearly laid-out roadmap toward a realistic goal. Notwithstanding such clever rhetorical orientations and narratives of future spaces for interactions, the current status of the metaverse cannot fully guarantee coherence and safety. These rhetorical approaches deliver hopes and dreams instead of practicalities and truths (Bibri, 2022). Under the circumstances of technologies struggling to grant access to educational development for all, there also needs to be a firm conclusion on what impact the continuing advancement of the metaverse will have (Zhai et al., 2022).

To be certain, the advancement of the metaverse has come a long way, making the distinction between physical and virtual increasingly blurry. As the transition of daily life into the digital follows its due course within many everyday spheres, more research and development needs to be conducted on how people can adapt to the many facets of the virtual environment, in order to excel as well as many people have on the internet (Lee et al., 2021). For example, academics should investigate ways that would allow the metaverse to transcend a single vision of the future (a thought paradigm usually proposed by big-tech leaders). As addressed in Chap. 9, more efforts should be made to open doors to diverse and plural representations within our profuse communicative worlds. The virtual platforms afforded by the metaverse should go in directions different from a single vision of the future (Bibri, 2022). Higher achievement in the metaverse will be possible if its designers endeavor to adapt the right mechanisms of social interaction, emotional attachment, and well-being (Lavdas et al., 2023).

Regulation of technology is a chief concern for scholars, developers, policymakers, and the public at large. To this point, the Collingridge dilemma is a theoretical model that can explain the difficulties that technology development faces: When change is effortless, the need for it is not part of any heated debate; on the other hand, when the need for change is evident, it becomes important, costly, difficult, and time-consuming (Collingridge, 1982). In its initial developments, the regulation of technology is a top priority (despite the fact that it is difficult to predict its impact). In turn, because it is difficult to determine what direction technology can take, controlling it is not an easy task either. In its subsequent developments, technology has shown to have clearer impacts, giving people a clearer indication as to what direction the technology has taken. Having said that, it can still be difficult to control it because it is deeply embedded within society (i.e., by being so entwined within material, economic, and social structures) (Kafaee et al., 2022). The Collingridge dilemma also posits that a developed technology can become an alternative in history. More precisely, technology development can be so influenced by historical conditions that will, in turn, shape historical events to come. It would, inter alia, affect new choices in daily life or even affect the shaping of other technologies (Liebert & Schmidt, 2010).

No matter what, it is our hope that emerging forms of communication can facilitate new critical, socio-technical approaches to research. The utmost significance of communication in the metaverse will continue to increase and offer invaluable solutions during present and future interpersonal, social, collaborative, educational, intercultural, health-related, and even individual or identity-related situations. Developing adequate skills and training constitutes an essential domain in which the metaverse can have positive applications and implications. With respect to research itself, with the rapid advancements in VR methodologies, it is high time to tackle issues within the metaverse as they will impact the aforementioned domains of communication.

References

Bibri, S. E. (2018). Approaches to futures studies: A scholarly and planning approach to strategic smart sustainable city development. In S. E. Bibri (Ed.), *Smart sustainable cities of the future: The untapped potential of big data analytics and context–aware computing for advancing sustainability* (pp. 601–660). Springer International.

Bibri, S. E. (2022). The social shaping of the metaverse as an alternative to the imaginaries of data-driven smart cities: A study in science, technology, and society. *Smart Cities, 5*(3), 832–874. https://doi.org/10.3390/smartcities5030043

Bibri, S. E., & Krogstie, J. (2020.): Article 5). The emerging data–driven smart city and its innovative applied solutions for sustainability: The cases of London and Barcelona. *Energy Inform, 3*. https://doi.org/10.1186/s42162-020-00108-6

Blackwell, L., Ellison, N., Elliott-Deflo, N., & Schwartz, R. (2019). Harassment in social virtual reality: Challenges for platform governance. *Proceedings of the ACM Human-Computer Interaction, 3*(100), 1–25. https://doi.org/10.1145/3359202

Bojić, L. (2022a). *Culture organism or techno-feudalism: How growing addictions and artificial intelligence shape contemporary society*. Institute for Philosophy and Social Theory.

Bojić, L. (2022b.): Article 22). Metaverse through the prism of power and addiction: What will happen when the virtual world becomes more attractive than reality? *European Journal of Futures Research, 10*(1). https://doi.org/10.1186/s40309-022-00208-4

Bougena, P. D., & O'Malley, P. (2009). Bureaucracy, imagination and U.S. domestic security policy. *Security Journal, 22*, 101–118. https://doi.org/10.1057/palgrave.sj.8350078

Brown, J. G., Bailenson, J. N., & Hancock, J. (2023). Misinformation in virtual reality. *Journal of Online Trust and Safety, 1*(5), 1–30. https://doi.org/10.54501/jots.v1i5.120

Caragnano, R. (2023). Labour law: New workplaces in the metaverse and opportunities for cultural and heritage professions. *Athens Journal of Law, 9*(2), 211–228.

Collingridge, D. (1982). *The social control of technology*. St. Martin's Press.

Croteau, D., & Hoynes, W. (2019). *Media/society: Technology, industries, content, and users* (6th ed.). Sage.

Davidson, K. J. (2022). *Semiotic modelling of identity and communication in virtual reality, augmented reality, and mixed reality*. University of Tartu.

Dolata, M., & Schwabe, G. (2023). What is the metaverse and who seeks to define it? Mapping the site of social construction. *Journal of Information Technology, 38*(3), 239–266. https://doi.org/10.1177/02683962231159927

Egliston, B., & Carter, M. (2021). Critical questions for Facebook's virtual reality: Data, power and the Metaverse. *Internet Policy Review, 10*(4), 1–23. https://doi.org/10.14763/2021.4.1610

Frohlich, M. (2022). Communicating in the metaverse? Keep your eyes wide open. *Provoke Media*. (January 18, 2022): Paragraph 7. Retrieved on October 25, 2022 from https://www.provokemedia.com/latest/article/communicating-in-the-metaverse-keep-your-eyes-wide-open

Hardman, S., Steinberger-Wilckens, R., & van der Horst, D. (2013). Disruptive innovations: The case for hydrogen fuel cells and battery electric vehicles. *International Journal of Hydrogen Energy, 38*(35), 15438–15451. https://doi.org/10.1016/j.ijhydene.2013.09.088

Horn, M. B., & Staker, H. (2014). *Blended: Using disruptive innovation to improve schools*. Jossey-Bass.

Howcroft, D., & Taylor, P. (2023). Automation and the future of work: a social shaping of technology approach. *New Technology, Work and Employment*. https://doi.org/10.1111/ntwe.12240

Kafaee, M., Daviran, E., & Taqavi, M. (2022). The QWERTY keyboard from the perspective of the Collingridge dilemma: Lessons for co-construction of human-technology. *AI & SOCIETY*, 1–13. https://doi.org/10.1007/s00146-022-01573-1

Lavdas, A. A., Mehaffy, M. W., & Salingaros, N. A. (2023.): Article 8). AI, the beauty of places, and the metaverse: Beyond 'geometrical fundamentalism'. *Architectural Intelligence, 2*. https://doi.org/10.1007/s44223-023-00026-z

Lee, L.-H., Braud, T., Zhou, P., Wang, L., Dianlei, X., Lin, Z., Kumar, A., Bermejo, C., & Hui, P. (2021). All one needs to know about metaverse: A complete survey on technological singularity, virtual ecosystem, and research agenda. *Journal of Latex Class Files, 14*(8), 1–66.

Leone, M. (2023). The spiral of digital falsehood in deepfakes. *International Journal for the Semiotics of Law, 36*(2), 385–405. https://doi.org/10.1007/s11196-023-09970-5

Liebert, W., & Schmidt, J. C. (2010). Collingridge's dilemma and technoscience. *Poiesis & Praxis, 7*(1), 55–71. https://doi.org/10.1007/s10202-010-0078-2

Liu, Z., & Zheng, Y. (2022). *AI ethics and governance: Black mirror and order*. Springer.

Loftus, E. F. (2005). Planting misinformation in the human mind: A 30-year investigation of the malleability of memory. *Learning Memory, 12*(4), 361–366. https://doi.org/10.1101/lm.94705

Lythreatis, S., Singh, S. K., & El-Kassar, A.-N. (2022.): Article 121359). The digital divide: A review and future research agenda. *Technological Forecasting and Social Change, 175*(6). https://doi.org/10.1016/j.techfore.2021.121359

MacKenzie, D. A., & Wajcman, J. (1999). *The social shaping of technology* (2nd ed.). McGraw Hill.

Maloney, D. (2021). *A youthful metaverse: Towards designing safe, equitable, and emotionally fulfilling social virtual reality spaces for younger users*. Clemson University.

Maloney, D., Freeman, G., & Robb, A. (2020a). It is complicated: Interacting with children in social virtual reality. *2020 IEEE Conference on Virtual Reality and 3D User Interfaces Abstracts and Workshops*, 343–347. https://doi.org/10.1109/VRW50115.2020.00075

Maloney, D., Freeman, G., & Robb, A. (2020b). A virtual space for all: Exploring children's experience in social virtual reality. *Proceedings of the Annual Symposium on Computer-Human Interaction in Play*, 472–483. https://doi.org/10.1145/3410404.3414268

Mansell, R. (2012). *Imagining the internet: Communication, innovation, and governance*. Oxford University Press.

Matusitz, J. (2022). *Fundamentals of public communication campaigns*. Wiley.

Matusitz, J., & Musambira, G. (2013). Power distance, uncertainty avoidance, and technology: Analyzing Hofstede's dimensions and human development indicators. *Journal of Technology in Human Services, 31*(1), 42–60. https://doi.org/10.1080/15228835.2012.738561

Murray, S. (2022). Meta employee says horizon worlds is 'sad' and 'empty,' player count falls far short of targets, *The Gamer* (October 16, 2022). Retrieved on March 3, 2023 from https://www.thegamer.com/meta-horizon-worlds-sad-empty/

Musambira, G., & Matusitz, J. (2015). Communication technology and culture: Analysing selected cultural dimensions and human development indicators. *International Journal of Technology Management & Sustainable Development, 14*(1), 17–28. https://doi.org/10.1386/tmsd.14.1.17_1

Nalbant, K. G., & Aydin, S. (2023). Development and transformation in digital marketing and branding with artificial intelligence and digital technologies dynamics in the metaverse universe. *Journal of Metaverse, 3*(1), 9–18. https://doi.org/10.57019/jmv.1148015

Paananen, V., Kiarostami, M. S., Lee, L.-H., Braud, T., & Hosio, S. (2022). From digital media to empathic reality: A systematic review of empathy research in extended reality environments. *arXiv*. Retrieved on October 25, 2022 from https://arxiv.org/pdf/2203.01375.pdf

Ragnedda, M., & Divide, T. T. D. (2017). *A weberian approach to digital inequalities*. Routledge.

Repez, F., & Popescu, M.-M. (2020). Social media and the threats against human security against human security deepfake and fake news. *Romanian Military Thinking, 4*, 44–55.

Rheingold, H. (1991). *Virtual reality*. Summit Books.

Simi, D., & Matusitz, J. (2022). Improving cultural diversity in U.S. higher education: General perspectives. *International Journal of Social Sciences & Educational Studies, 9*(4), 117–129. https://doi.org/10.23918/ijsses.v9i4p117

Vasist, P. N., & Krishnan, S. (2023). Engaging with deepfakes: A meta-synthesis from the perspective of social shaping of technology theory. *Internet Research, 33*(5), 1670–1726. https://doi.org/10.1108/INTR-06-2022-0465

Wang, Y., & Wei, D. (2022). Evolution logic and path choice of IP development of media virtual image in the digital twin era. *Meta, 3*(2), 1–8. https://doi.org/10.54517/met.v3i2.2131

Williams, R., & Edge, D. (1996). The social shaping of technology. *Research Policy, 25*(6), 865–899. https://doi.org/10.1016/0048-7333(96)00885-2

Winner, L. (1980). Do artifacts have politics? *Daedalus, 109*(1), 121–136.

Zhai, X., Chu, X., Wang, M., Zhang, Z., & Dong, Y. (2022). Education Metaverse: Innovations and challenges of the new generation of internet education formats. *Meta, 3*(1), 1–13. https://doi.org/10.54517/met.v3i1.1804

Zhang, M., Wang, Y., Zhou, J., & Pan, Z. (2022). SimuMan: A simultaneous real-time method for representing motions and emotions of virtual human in metaverse. In *Internet of Things—ICIOT 2021* (pp. 77–89). Springer.

Index

J. Matusitz, J. Dacas, *Communication in the Metaverse*, Signals and Communication Technology, https://doi.org/10.1007/978-3-031-63485-7

The manufacturer's authorised representative in the EU is Springer Nature Customer Service Centre GmbH, Europaplatz 3, 69115 Heidelberg, Germany. If you have any concerns regarding our products, please contact ProductSafety@springernature.com

Printed and bound by CPI Group (UK) Ltd, Croydon, CR0 4YY
30/09/2026
02225592-0003